**FIFTH EDITION**

# COMMERCIAL PILOT SYLLABUS

by

**Irvin N. Gleim, Ph.D., CFII**

and

**Garrett W. Gleim, CFII**

## ABOUT THE AUTHORS

Irvin N. Gleim earned his private pilot certificate in 1965 from the Institute of Aviation at the University of Illinois, where he subsequently received his Ph.D. He is a commercial pilot and flight instructor (instrument) with multi-engine and seaplane ratings, and is a member of the Aircraft Owners and Pilots Association, American Bonanza Society, Civil Air Patrol, Experimental Aircraft Association, National Association of Flight Instructors, and Seaplane Pilots Association. Dr. Gleim is the author of flight maneuvers and practical test prep books for the sport, private, instrument, commercial, and flight instructor certificates/ratings and is the author of study guides for the sport, private/recreational, instrument, commercial, flight/ground instructor, fundamentals of instructing, airline transport pilot, and flight engineer FAA knowledge tests. Three additional pilot training books are *Pilot Handbook*, *Aviation Weather and Weather Services*, and *FAR/AIM*.

Dr. Gleim has also written articles for professional accounting and business law journals, and is the author of widely used review manuals for the CIA (Certified Internal Auditor) exam, the CMA (Certified Management Accountant) exam, the CPA (Certified Public Accountant) exam, and the EA (IRS Enrolled Agent) exam. He is Professor Emeritus, Fisher School of Accounting, University of Florida, and is a CFM, CIA, CMA, and CPA.

Garrett W. Gleim earned his private pilot certificate in 1997 in a Piper Super Cub. He is a commercial pilot (single- and multi-engine), ground instructor (advanced and instrument), and flight instructor (instrument and multi-engine), and is a member of the Aircraft Owners and Pilots Association and the National Association of Flight Instructors. Mr. Gleim is the author of study guides for the sport, private/recreational, instrument, commercial, flight/ground instructor, fundamentals of instructing, and airline transport pilot FAA knowledge tests. He received a Bachelor of Science in Economics from The Wharton School, University of Pennsylvania. Mr. Gleim is also a CPA (not in public practice).

**Gleim Publications, Inc.**
P.O. Box 12848 · University Station
Gainesville, Florida 32604
(352) 375-0772
(800) 87-GLEIM or (800) 874-5346
Fax: (352) 375-6940
Internet: www.gleim.com
Email: admin@gleim.com

For updates to the second printing of the fifth edition of
***Commercial Pilot Syllabus***

**Go To:** www.gleim.com/updates

**Or:** Email update@gleim.com with **CPSYL 5-2** in the subject line. You will receive our current update as a reply.

Updates are available until the next edition is published.

ISSN 1528-6304
ISBN 978-1-58194-973-5



First Printing: June 2012
Second Printing: September 2014



**ACKNOWLEDGMENTS**

The airplane image on the front cover is a Piper Seminole, courtesy of Piper Aircraft, Inc.

**HELP!!**

This is one book in a flight training series of six books, listed below, designed specifically for persons who aspire to earn a commercial pilot certificate.

*Commercial Pilot Syllabus*
*Commercial Pilot Flight Maneuvers and Practical Test Prep*
*Commercial Pilot FAA Knowledge Test*
*Aviation Weather and Weather Services*
*Pilot Handbook*
*FAR/AIM*

Please send any corrections and suggestions for subsequent editions to the authors, c/o Gleim Publications, Inc. The last page in this book has been reserved for you to make comments and suggestions. It can be torn out and mailed or faxed. Alternatively, you may email your comments and suggestions to aviation@gleim.com.

Also, please bring this book to the attention of flight instructors, fixed-base operators, and others interested in flying. Wide distribution of this series of books and increased interest in flying depend on your assistance and good word. Thank you.

**Environmental Statement** -- This book is printed on high-quality, environmentally friendly groundwood paper, sourced from certified sustainable forests and produced either TCF (totally chlorine-free) or ECF (elementally chlorine-free). Our recyclable paper is more porous than coated paper, so we recommend marking it with a non-bleed-through highlighter.

**NOTE: UPDATES**

Visit our website or email update@gleim.com as described at the top right of this page for the latest updates and information on all of our products. Updates for this edition will be available until the next edition is published. To continue providing our customers with first-rate service, we request that questions about our materials be sent to us via email to aviation@gleim.com. The appropriate staff member will give each question thorough consideration and a prompt response. Questions concerning orders, prices, shipments, or payments will be handled via telephone by our competent and courteous customer service staff.

# TABLE OF CONTENTS



***IF FOUND, PLEASE CONTACT***

Pilot Name ______________________________

Address ______________________________

______________________________

______________________________

Telephone # ______________________________

Email ______________________________

# PREFACE

Thank you for choosing GLEIM. Our training materials (books, software, and online) are intuitively appealing and thus very effective in transferring knowledge to you. The Gleim system saves you time, money, and frustration vs. other aviation training programs.

This syllabus will facilitate your studies and training for your commercial pilot certificate and, optionally, for your multi-engine rating as well.

1. Please read the following Introduction carefully.
2. The "Lesson Sequence and Times" section shows ground lessons being completed as you complete flight lessons. We encourage you to work ahead in your ground lessons and begin them (and even complete them) prior to beginning your flight training.
3. The OBJECTIVE is to develop "PTS level" proficiency as quickly as possible. "PTS level" means that you can perform at the level required by the FAA's Practical Test Standards (PTS).
4. Homework consists of pages to read and/or study in your Gleim ***Commercial Pilot FAA Knowledge Test*** book, Gleim ***Commercial Pilot Flight Maneuvers and Practical Test Prep*** book, Gleim ***Pilot Handbook***, Gleim ***Aviation Weather and Weather Services*** book, Gleim ***FAR/AIM*** book, and the Pilot's Operating Handbook or Operating Limitations for your airplane. Each flight lesson also directs you to review topics and material studied for previous lessons.
   a. The Commercial Pilot Kit does not contain ***Pilot Handbook*** or ***FAR/AIM*** since these books were included in the Gleim Private Pilot Kit. Additionally, ***Aviation Weather and Weather Services*** is not included in the Commercial Pilot Kit since this book was included in the Gleim Instrument Pilot Kit. If you do not have them, please order your copies today.
   b. Further, the online Multi-Engine Add-On Rating Course, used as ground lessons in Part II of this syllabus, is not included with the Commercial Pilot Kit since pilots will pursue the training after their initial commercial certification is complete. Pilots who want to pursue multi-engine certification can purchase the online course when their training actually begins.

Why is the GLEIM SYSTEM different? It focuses on successful completion, as quickly and as easily as possible. The requirements for earning your commercial pilot certificate (single-engine-land) are listed beginning on page 2. Requirements for the multi-engine add-on rating can be found on page 3. This syllabus facilitates your flight training so you achieve a "PTS level" of proficiency on the 46 FAA tasks required for initial commercial pilot certification as well as the 26 tasks required as part of the multi-engine add-on as quickly as possible!

GO FOR IT! Start studying for your FAA knowledge test today. Refer to Study Units 1 through 11 in ***Commercial Pilot FAA Knowledge Test*** and associated textbooks. Start studying for your FAA practical test by reading Study Unit 3 (of Part I) in ***Commercial Pilot Flight Maneuvers and Practical Test Prep***.

We have an easy-to-follow and easy-to-complete study system. From the very start, we want you to focus on success. This means answering over 80% of the FAA knowledge test questions correctly AND being able to explain and demonstrate all appropriate FAA practical test tasks to your CFI at a "PTS level" proficiency.

Enjoy Flying Safely!

*Irvin N. Gleim*
*Garrett W. Gleim*
September 2014

# INTRODUCTION



This syllabus is a step-by-step lesson plan for your commercial pilot training. It is intended to be used in conjunction with the following five Gleim books:

***Commercial Pilot FAA Knowledge Test***
***Commercial Pilot Flight Maneuvers and Practical Test Prep***
***Aviation Weather and Weather Services***
***Pilot Handbook***
***FAR/AIM***

NOTE: ***Pilot Handbook*** and ***FAR/AIM*** are not included in the Commercial Pilot Kit since these books were included in the Gleim Private Pilot Kit. Also, ***Aviation Weather and Weather Services*** is not included in the Commercial Pilot Kit since this book was included in the Gleim Instrument Pilot Kit. They are available at your local FBO, flight school, bookstore, etc., or directly from Gleim at (800) 874-5346 or www.gleim.com (see product listing at the back of this book). Finally, the online Multi-Engine Add-On Rating Course, used as ground lessons in Part II of this syllabus, is not included with the Commercial Pilot Kit since pilots will pursue the training after their initial commercial certification is complete. Pilots who want to pursue multi-engine certification can purchase the course when their training actually begins.

Your flight instructor (or flight school) will retain a record of your progress through each lesson in this book using the Commercial Pilot Training Record, included in pilot kits or sold separately and discussed in greater detail on page 4.

## WHAT ELSE DO YOU NEED?

If you purchased this syllabus as part of the Gleim **Commercial Pilot Kit**, you will need to purchase (1) a local sectional chart and (2) an *Airport/Facility Directory (A/FD)*. They are published at intervals of 6 months and 56 days, respectively. You will need a current copy of each for your FAA practical test. Gleim does not include these publications in your kit because there are 37 different sectional charts and seven different *A/FDs* for the conterminous U.S.

Additionally, you will need to purchase a copy of the Pilot's Operating Handbook (POH) (sometimes called an Information Manual) or Operating Limitations for the make and model of your training airplane. Alternatively, you may make a photocopy if a POH is not available for purchase.

## REQUIREMENTS FOR A COMMERCIAL PILOT CERTIFICATE – SINGLE-ENGINE LAND

You must meet a number of requirements to earn your commercial pilot certificate. The final step is your FAA practical test, which will be conducted by an FAA-designated pilot examiner or an FAA inspector, unless your Part 141 pilot school has examining authority to conduct the practical test. Your practical test will consist of an approximately 1- to 2-hour oral exam followed by a 1- to 2-hour flight test. You will be well prepared for your practical test by your CFI and your Gleim pilot training materials. In addition, you must meet the following requirements:

1. Be at least 18 years of age and hold at least a private pilot certificate with an airplane category rating and single-engine class rating.
2. Be able to read, write, and converse fluently in English (certificates with operating limitations may be available for medical-related deficiencies).
3. Hold a current FAA medical certificate.
   a. Your medical exam will be conducted by an FAA-designated aviation medical examiner (AME).
   b. Ask your CFI or call your local flight school for the names and telephone numbers of the AMEs in your area, or visit www.gleim.com/aviation/amesearch.php for a listing of AMEs by state and city.
   c. Review Form 8500-8 and make sure you are able to fill out items 1 through 20 to reduce the amount of time at your AME's office.
      1) A reproduction of FAA Form 8500-8, *Application for Airman Medical Certificate or Airman Medical and Student Pilot Certificate*, is available for review on pages 167 through 169 of this syllabus.
      2) Use the FAA MedXPress system to fill out your medical application before going to the AME's office. Access the system at medxpress.faa.gov.
4. Pass your FAA commercial pilot (airplane) airman knowledge test, which consists of 100 multiple-choice questions administered at an FAA-designated computer testing center. Everything you need to prepare for your FAA commercial pilot (airplane) airman knowledge test is in your Gleim ***Commercial Pilot FAA Knowledge Test***, ***Commercial Pilot Flight Maneuvers and Practical Test Prep***, ***FAR/AIM***, ***Pilot Handbook***, and ***Aviation Weather and Weather Services*** books. Gleim **FAA Test Prep Online** will facilitate your study. (To find out exactly what to expect from the computer testing center of your choice, use the **FAA Test Prep Online's** convenient vendor emulation testing format.) You may also use our convenient **Online Ground School** program to prepare for the knowledge test. Successful completion of the course will provide you with a knowledge test endorsement, which you will need to actually sit for the exam.
   a. We have estimated 35 hours for complete preparation for your commercial pilot airman knowledge test. See "Commercial Pilot Syllabus Lesson Sequence and Times."
   b. The further you study (not read) to prepare for your FAA commercial pilot airman knowledge test before you commence your flight lessons, the better!
      1) Under Part 141, you should be enrolled in the school's commercial pilot certification course before beginning your study program.
5. Undertake flight training as described in Part I, Lessons 1 through 60. Many of the lessons may require more than one flight to complete. We also have provided space for your instructor to record extra flights within each lesson as needed to make you comfortable and proficient.
   a. The importance of home study certainly applies to piloting. Do NOT show up at the airport expecting to be spoon-fed by your CFI. If you do, your flight training will be slow going and frustrating.

b. It's fun to be successful! Be overly prepared before you get to the airport for each flight lesson.

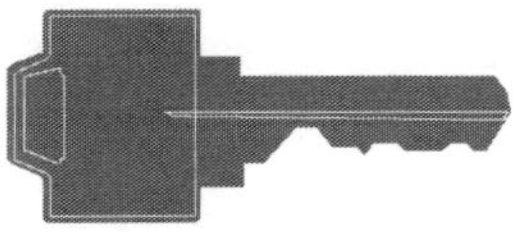

The KEY TO SUCCESS in your flight training, which also minimizes cost and frustration, is your study and preparation at home before flying with your flight instructor. The more you know about flying, flight training, and each flight lesson, the better you will do.

6. PASS your FAA practical test. See Study Unit 3 in ***Commercial Pilot Flight Maneuvers and Practical Test Prep***.
7. For additional detail regarding FAA commercial pilot requirements, see the introduction in the ***Commercial Pilot FAA Knowledge Test*** book or pages 2 and 3 in the ***Commercial Pilot Flight Maneuvers and Practical Test Prep*** book.

## REQUIREMENTS FOR A MULTI-ENGINE ADD-ON RATING FOR COMMERCIAL PILOTS

The vast majority of multi-engine pilots do not accomplish their initial pilot certification in a multi-engine airplane, but rather add the multi-engine rating onto their existing pilot certificate. The requirements below explain how a certificated commercial pilot would go about adding multi-engine privileges to his/her pilot certificate.

1. Hold a current and valid commercial pilot certificate as well as a current and valid FAA medical certificate.
2. Complete the ground and flight training specified in Part II of this syllabus.
   a. If you intend to train at a Part 141 approved school, you should enroll in its training program before beginning your study program.
3. Pass an FAA practical flight test on 26 specific PTS tasks.
   a. If you do not hold an instrument rating, you are only required to complete 25 tasks (i.e., you do not have to complete an instrument skills evaluation unless you hold an instrument rating on your pilot certificate).

## HOW TO PROCEED

1. If you are reading someone else's copy of ***Commercial Pilot Syllabus***, obtain a Gleim **Commercial Pilot Kit** from your FBO/flight school/bookstore, or for publisher-direct service, call (800) 874-5346 or visit www.gleim.com.

   The **Commercial Pilot Kit** contains ***Commercial Pilot FAA Knowledge Test*** book, **FAA Test Prep Online**, **Online Ground School**, ***Commercial Pilot Flight Maneuvers and Practical Test Prep***, and this ***Commercial Pilot Syllabus***.
2. Read this Introduction (9 pages).
3. Begin preparing for your FAA commercial pilot airman knowledge test by using your Gleim books, **FAA Test Prep Online**, and/or **Online Ground School**. See your ground training syllabus.
4. Select a CFI and/or flight school. See Study Unit 1 in ***Commercial Pilot Flight Maneuvers and Practical Test Prep***.
5. Begin your flight training, scheduling at least two lessons a week. Prepare thoroughly at home for each flight lesson. Use the flight training syllabus in this book.
6. After you receive your commercial pilot certificate, consult Part II of this book on how you can add a multi-engine rating to that certificate.

## PART 141 VS. PART 61 SCHOOLS

Title 14 of the Code of Federal Regulations (14 CFR), which is commonly referred to as the FARs by pilots, lists the requirements to obtain your commercial pilot certificate. Pilot (or flight) schools can conduct your training by following either 14 CFR Part 141 or 14 CFR Part 61.

An eligible pilot school may be issued a Part 141 pilot school certificate by the FAA after completion of an application process. Part 141 pilot schools are more regulated than a Part 61 pilot school. Additionally, a Part 141 pilot school is required to have its facilities and airplanes inspected, and it must have its ground and flight training syllabi approved by the FAA.

The majority of pilot schools, and flight instructors not associated with a Part 141 pilot school, provide the required training specified under Part 61.

The major difference between a Part 141 and a Part 61 pilot school is that a Part 141 pilot school requires FAA approval and can take a pilot through the school's private pilot, instrument rating, and commercial pilot courses in a minimum of 190 hr. compared to 250 hr. under Part 61. Additionally, Part 61 requires 100 hr. of pilot in command (PIC) time. Part 141 does not have a minimum PIC flight time requirement.

The Gleim syllabus has been reviewed by the FAA in Washington, D.C., and found to adequately meet the requirements of a syllabus under Part 141 or Part 61, as appropriate (see FAA letter on back cover). Thus, the Gleim ***Commercial Pilot Syllabus*** can be used by any Part 141 school with minimal effort. We suggest that school write a letter to its FAA inspector advising the FAA of the school's intent and the names of students who are going to be trained under this Gleim syllabus.

If a Part 141 pilot school cannot or will not use this syllabus, consider finding another Part 141 or Part 61 pilot school for your training OR please call (800) 874-5346 if you have questions or problems.

## EXPLANATION OF *COMMERCIAL PILOT TRAINING RECORD* BOOKLET

Also available separate from this syllabus is a *Commercial Pilot Training Record*. This Pilot Training Record is provided for flight schools that conduct training under 14 CFR Part 141, which requires that detailed training records be maintained for each student. When properly completed, the booklet will meet the training record requirements of Part 141.

### Training Record Elements

The training record booklet consists of three main sections:

- The front cover contains student personal information and information about the training course.
- The back cover serves as a ground training and student evaluation record.
- The inside of the booklet consists of a two-part flight training record and a separate record of instructor endorsements.

## Using the Training Record

**Front Cover:** The front cover of the training record should be filled out by the student, his/her flight instructor, and the chief flight instructor at the time of enrollment. Spaces provided to record credit awarded for previous ground and flight training should be completed by the chief instructor. The chief instructor should also complete the appropriate enrollment certificate (found on pages 159 and 161 of this syllabus) and place it in the pocket at the back of the training record.

At the completion of training, the chief instructor should complete the information on the front cover, as appropriate (e.g., graduation, transfer, or termination). If the student has graduated, the appropriate graduation certificate (found on pages 163 and 165 of this syllabus) should be completed and placed in the pocket at the back of the training record.

**Back Cover:** The ground training record should be filled out by the instructor after each ground lesson is completed, regardless of whether ground training is being conducted formally or as a self-study program. The time spent and date of completion should be noted, and the record initialed by both the student and the instructor.

The stage and end-of-course test records should be filled out by the instructor after each stage and end-of-course test has been taken by the student, graded, and reviewed with the instructor. The date of the test, the result, and the date of the review should be noted. The record should then be initialed by the student and signed by the instructor. Each stage and end-of-course exam answer sheet should also be placed in the pocket at the back of the training record.

The student evaluation records should be filled out by the check instructor following each stage check evaluation, and by the chief instructor after the end-of-course check. The date and result of the test should be noted, and the record initialed by the student and signed by the check instructor or chief instructor (the chief instructor must sign the record for the end-of-course check).

**Inside:** The flight training record consists of two parts: the Flight Record is a chronological record of each training flight that is made during the course, while the Flight Lesson Record is an itemized record of the student's performance on the lesson items listed in each specific flight lesson. Note that it is divided into two parts, like this syllabus, to make it easy to record your progress in the initial commercial pilot certification course as well as the multi-engine add-on rating course. The record of instructor endorsements is a record of information related to each flight instructor endorsement that is pertinent to the course of training.

The Flight Record should be used to record the following information for each flight that is performed during the course: date of the flight, aircraft make and model, aircraft registration number, flight lesson(s) undertaken during the flight (this item has been left for the instructor to complete in order to accommodate lessons that require multiple flights to complete or that must be completed out-of-sequence), initials of the instructor conducting the flight (if applicable), and the types and amounts of flight time accumulated during the flight and during the entire course to date.

The Flight Lesson Record should be used to record the student's performance on each lesson item that is contained in the lesson(s) that were undertaken during the flight. The Flight Lesson Record consists of 70 individual lesson records (one for each lesson contained in the syllabus – 60 lessons in Part I and 10 lessons in Part II). Each individual lesson record contains an itemized reprint of the lesson items contained in the corresponding flight lesson. Because it is recognized that many lessons take more than one flight to complete, each individual lesson record is designed to allow the instructor to record and grade the student's performance on each lesson item for up to eight flights.

The Flight Lesson Record should be used to record the following information for each flight lesson: the date of each flight undertaken to complete the lesson, the time spent during each pre- and post-flight briefing, and the student's performance on each lesson item. When the lesson is completed, the date of the lesson's completion should be noted, and the record initialed by the student and signed by the instructor who conducted the final flight for that lesson.

The record of instructor endorsements should be filled out by the instructor at the time that each logbook endorsement is provided. The date of the endorsement and any applicable conditions should be noted, and the record should be signed by the instructor. The instructor's flight instructor certificate number and expiration date should also be noted.

## USE OF A FLIGHT SIMULATOR, A FLIGHT TRAINING DEVICE, OR A PERSONAL COMPUTER-BASED AVIATION TRAINING DEVICE

While the Gleim ***Commercial Pilot Syllabus*** was designed to be used in a course of training using only an airplane, it may also be used in a course of training in which the use of a qualified and approved flight simulator or flight training device has been appropriately integrated with the approval of the jurisdictional FSDO.

Under 14 CFR section 61.4, Qualification and Approval of Flight Simulators and Flight Training Devices, a Part 141 pilot school may elect to use either a flight simulator or a flight training device in accordance with provisions of section 61.4(a). The Part 141 pilot school may also elect to use a flight training device as described under section 61.4(b). However, the creditable use of a flight simulator or flight training device identified under section 61.4(a) in Part 141 appendices is governed by 14 CFR section 141.41(a), and the creditable use of a flight training device identified under section 61.4(b) is governed by 14 CFR section 141.41(b).

A pilot school operating under Part 61 may also elect to use the flight simulators and flight training devices identified above in accordance with specific authorization for that use as outlined in the various sections of Part 61 [such as a maximum of 50 hours of flight training under section 61.129(i)]. This creditable use should be dedicated to instrument procedures and associated flight maneuver review. (See Commercial Pilot Practical Test Standards, FAA-S-8081-12C, "Use of FAA-Approved Flight Simulation Training Device (FSTD)," for guidance concerning the use of these devices.)

## AIRPLANE(S) AND LOCAL AIRPORT(S) WORKSHEETS

The airplane(s) worksheets on pages 7 and 8 are to be used as a study reference of fuel, oil, and airspeeds for the airplane(s) you may use during your training. The first airplane worksheet is designed to accommodate single-engine airplanes, and the second worksheet is intended for multi-engine airplanes. The local airport(s) worksheet on page 9 is to be used as a study reference of radio frequencies, traffic pattern direction, and runway lengths at your primary airport and any other local airports you may use during your training.

# SINGLE-ENGINE AIRPLANE(S) WORKSHEET

| | | | | | |
|---|---|---|---|---|---|
| N-number | | | | | |
| Make & Model | | | | | |
| Max. Ramp Wt. | | | | | |
| Fuel Capacity | | | | | |
| Min. Fuel for Flight | | | | | |
| Oil Capacity | | | | | |
| Min. Oil for Flight | | | | | |
| $V_{SO}$ | | | | | |
| $V_{S1}$ | | | | | |
| $V_R$ | | | | | |
| $V_X$ | | | | | |
| $V_Y$ | | | | | |
| $V_{FE}$ | | | | | |
| $V_A$ | | | | | |
| $V_{LO}$ | | | | | |
| $V_{LE}$ | | | | | |
| $V_{NO}$ | | | | | |
| $V_{NE}$ | | | | | |
| Best Glide | | | | | |

# MULTI-ENGINE AIRPLANE(S) WORKSHEET

| | | | | | |
|---|---|---|---|---|---|
| N-number | | | | | |
| Make & Model | | | | | |
| Max. Ramp Wt. | | | | | |
| Fuel Capacity | | | | | |
| Min. Fuel for Flight | | | | | |
| Oil Capacity | | | | | |
| Min. Oil for Flight | | | | | |
| $V_{SO}$ | | | | | |
| $V_{S1}$ | | | | | |
| $V_{MC}$ | | | | | |
| $V_R$ | | | | | |
| $V_X$ | | | | | |
| $V_{XSE}$ | | | | | |
| $V_Y$ | | | | | |
| $V_{YSE}$ | | | | | |
| $V_{FE}$ | | | | | |
| $V_A$ | | | | | |
| $V_{LO}$ | | | | | |
| $V_{LE}$ | | | | | |
| $V_{NO}$ | | | | | |
| $V_{NE}$ | | | | | |
| Best Glide | | | | | |

# LOCAL AIRPORT(S) WORKSHEET

| | | | | | |
|---|---|---|---|---|---|
| Airport Name | | | | | |
| Identifier | | | | | |
| Elevation | | | | | |
| ATIS | | | | | |
| Ground | | | | | |
| Tower | | | | | |
| UNICOM | | | | | |
| Runway | | | | | |
| Length | | | | | |
| Traffic Pattern | Left or Right | Left or Right | Left or Right | Left or Right | Left or Right |
| Obstructions | | | | | |
| Runway | | | | | |
| Length | | | | | |
| Traffic Pattern | Left or Right | Left or Right | Left or Right | Left or Right | Left or Right |
| Obstructions | | | | | |
| Runway | | | | | |
| Length | | | | | |
| Traffic Pattern | Left or Right | Left or Right | Left or Right | Left or Right | Left or Right |
| Obstructions | | | | | |
| Runway | | | | | |
| Length | | | | | |
| Traffic Pattern | Left or Right | Left or Right | Left or Right | Left or Right | Left or Right |
| Obstructions | | | | | |
| Traffic Pattern Altitude | | | | | |

# Update Service

Visit the **GLEIM**® website for free updates,
which are available until the next edition is published.

# gleim.com/updates

# PART I: COMMERCIAL PILOT TRAINING SYLLABUS AIRPLANE SINGLE-ENGINE LAND



## PART 141 STUDENT INFORMATION

### Enrollment Prerequisites

You must hold at least a private pilot certificate, with an airplane category and single-engine land rating and an instrument rating, with an airplane category, prior to enrolling in the flight portion of the commercial pilot certification course. If you do not hold an instrument rating at the time of enrollment, you may continue the course if you are concurrently enrolled in an approved instrument rating course.

### Graduation Requirements

You must complete the training specified in this syllabus, with a minimum of 35 hours of ground training in the specified aeronautical knowledge areas and a minimum of 120 hours of flight training. These requirements are reflected in the Gleim ground and flight training syllabus.

### Stage Checks

You must score a minimum of 80% on the knowledge test at the completion of each stage in the ground training syllabus and on the comprehensive end-of-course knowledge test at the conclusion of the training.

You must satisfactorily complete a stage check at the completion of each stage of the flight training syllabus, as well as a comprehensive end-of-course check when all flight training is complete.

### Credit for Previous Training

You may be given credit toward this commercial pilot course for previous pilot experience and knowledge [14 CFR 141.77(c)]:

1. If the credit is based on a Part 141 training course, the credit may be 50% of the requirements for this course.
2. If the credit is based on a Part 61 course, the credit cannot exceed 25% of the requirements for this course.

The receiving school will determine the amount of course credit to be given, based on a proficiency test, a knowledge test, or both.

## STRUCTURE OF THE GLEIM COMMERCIAL PILOT SYLLABUS (SINGLE-ENGINE LAND)

This syllabus consists of a ground training syllabus and a flight training syllabus. The ground and flight training may be done together as an integrated course of instruction, or each may be done separately. If done separately, the ground syllabus may be conducted as a home-study course or as a formal ground school.

This syllabus was constructed using the building-block progression of learning, in which the student is required to perform each simple task correctly before a more complex task is introduced. This method will promote the formation of correct habit patterns from the beginning.

### Ground Training Syllabus

The ground training syllabus contains 11 lessons, which are divided into two stages. The ground training syllabus meets the training requirements of Appendix D to Part 141 and Sec. 61.125. The ground training can be conducted concurrently with the flight training, with the ground lessons completed in order as outlined in the lesson matrix. Ground training may also be conducted as part of a formal ground school or as a home-study program.

It is recommended that the lessons be completed in sequence, but the syllabus is flexible enough to meet the needs of an individual student or of a particular training environment. When departing from the sequence, the instructor is responsible for considering the blocks of learning affected and, if used by a Part 141 pilot school, whether it would affect FAA approval.

Each ground lesson involves studying the appropriate section(s) in Gleim ***Aviation Weather and Weather Services***, ***FAR/AIM,*** and ***Pilot Handbook***. After each reading assignment is completed, you answer the questions in the appropriate study unit in Gleim ***Commercial Pilot FAA Knowledge Test*** book and review incorrect responses with your instructor.

Alternatively, Gleim **FAA Test Prep Online** or **Online Ground School** can be used to answer the questions at the end of each ground lesson. Our software and online course contain the FAA figures and outlines in addition to the questions.

**FAA Test Prep Online** allows you to select either STUDY MODE or TEST MODE. In STUDY MODE, the software provides you with an explanation of each answer you choose (correct or incorrect).

In TEST MODE, you decide the format -- CATS, PSI/LaserGrade, or Gleim. The software can emulate the operation of the FAA-approved computer testing companies. Thus, you will have a complete understanding of exactly how to take an FAA pilot knowledge test before you go to a computer testing center. When you finish your test, you can study the questions missed and access answer explanations.

At the end of each stage, you are required to complete the stage knowledge test before proceeding to the next stage. The end-of-course knowledge test is completed after the Stage Two knowledge test. Shortly after the end-of-course test, you should take the FAA commercial pilot (airplane) knowledge test. The stage knowledge tests in the ground syllabus will refer you to FAA figures located after the knowledge tests in this book. If you utilize the **Online Ground School**, you will be provided stage tests and an end-of-course test online. Upon successful completion of the end-of-course knowledge test, the system will automatically generate an endorsement for you to take the knowledge test. Online Ground School is especially valuable for a student who is studying for the knowledge test before beginning his/her flight training.

If this ground training is used as home study, we recommend that you complete the syllabus as quickly as possible and pass the FAA commercial pilot (airplane) knowledge test so you will have more time to prepare for your flight lessons.

## Flight Training Syllabus

The flight training syllabus contains 60 lessons, which are divided into three stages. It is recommended that each lesson be completed in sequential order.

Stage One of the flight training syllabus is designed to provide you with the knowledge and the skill required to safely operate a complex airplane. Additionally, you will be introduced to commercial flight maneuvers.

Stage Two is designed to increase your skill and proficiency in both day and night VFR cross-country operations.

Stage Three is designed to provide you with the instruction and practice necessary to attain the skill level required to pass the FAA commercial pilot practical test.

**Stage checks.** Stage checks are designed to ensure that the student has acquired the necessary knowledge and skill. The Stage One check (Lesson 17) is to ensure that you are proficient in operating a complex airplane. The Stage Two check (Lesson 34) is to ensure that you are proficient in planning and conducting VFR cross-country flights. The Stage Three check (Lesson 59) is to ensure that you are proficient in commercial pilot flight maneuvers. The End-of-Course check (Lesson 60) is to ensure that you are proficient in all commercial pilot knowledge and skill areas and are ready for the commercial pilot practical test. It is also the final evaluation for a Part 141 graduation certificate.

The chief flight instructor (Part 141) is responsible for ensuring that each pilot accomplishes the required stage checks and the end-of-course test. The chief flight instructor may delegate authority for conducting stage checks and the end-of-course test to the assistant chief flight instructor or a check instructor.

Stage checks will be used as a review by instructors training under Part 61 to ensure that the student has the appropriate knowledge and skills.

**Sequence of a flight lesson.** Each flight lesson will begin with a preflight briefing. During this time, the instructor should first answer any questions you may have from the previous lesson. Next, your instructor will brief you on the lesson content, followed by an evaluation of your preparation for the lesson.

During the flight portion of the lesson, your instructor should begin with those maneuvers listed as "review items" before introducing new maneuvers. The time required for each lesson will vary depending on the airport and the location of the training areas.

At the end of each lesson, your instructor will conduct a postflight critique and a preview of the next lesson. This time should be used to review the good points during the lesson, to identify and explain fully any problem areas, and to discuss how to correct the problems.

The length of the preflight briefing and postflight critique will vary with each student and with his/her degree of preparedness for the lesson.

**Pilot preparation.** The key to minimizing frustration and costs is preparation. You should budget an average of 2 to 4 hours of home study prior to each flight lesson. Learning will be easier when you are fully prepared so that your instructor can maximize the time spent in flight training.

**Use of a complex airplane.** A complex airplane is defined as an airplane that has retractable landing gear, flaps, and a controllable pitch propeller (including airplanes where the propeller pitch is digitally controlled, such as a full authority digital engine control – FADEC). During the commercial pilot practical test, the takeoff and landing maneuvers, and appropriate emergency procedures, must be accomplished in a complex airplane. We have developed the flight training syllabus under the assumption that a pilot will use a complex airplane for the entire practical test. If two airplanes are to be used during the practical test, we suggest that Flight Lessons 51, 52, and 54 remain as lessons using a complex airplane.

## COMMERCIAL PILOT SYLLABUS LESSON SEQUENCE AND TIMES

The Ground Syllabus follows on pages 17 through 32, and the Flight Syllabus follows on pages 33 through 100.

The table below and on the following pages lists the sequence of the flight and ground lessons and the minimum time for each lesson. The times listed are for instructor/student guidance only and are not meant to be mandatory times. These times will ensure that the minimum time requirements for aeronautical knowledge and flight training are in compliance with Part 141, Appendix D, Commercial Pilot Certification Course. See the notes following the table for additional Part 61 training requirements.

The major difference between a Part 141 and a Part 61 flight school is the minimum amount of flight time required to obtain a commercial pilot certificate. A Part 141 flight school can take a pilot through its private pilot, instrument rating, and commercial pilot courses in a minimum of 190 hr. compared to 250 hr. under Part 61. Additionally, Part 61 requires 100 hr. of pilot in command (PIC) and 50 hours of cross-country time.

Each training flight must include a preflight briefing and a postflight critique of the pilot's performance by the instructor.

| | | Flight | | | | | | | | Ground |
|---|---|---|---|---|---|---|---|---|---|---|
| LESSON | Page | Flight Training Dual | Flight Training Solo | Dual Cross-Country | Solo Cross-Country | Instrument | Night Dual | Night Solo | Complex Airplane Dual | Aeronautical Knowledge Training |
| **STAGE ONE** | | | | | | | | | | |
| Flight 1 Traffic Pattern Operations | 39 | 1.5 | | | | | | | | |
| Ground 1 Airplanes/Aerodynamics | 20 | | | | | | | | | 3.0 |
| Flight 2 Slow Flight/Stalls/Emergency | 40 | 1.5 | | | | 0.5 | | | | |
| Flight 3 Solo Practice | 41 | | 1.5 | | | | | | | |
| Flight 4 Maneuvers Review | 42 | 1.5 | | | | 0.5 | | | | |
| Ground 2 Airplane Instruments, Engines, and Systems | 21 | | | | | | | | | 3.0 |
| Flight 5 Intro to Complex Airplanes | 43 | 1.0 | | | | | | | 1.0 | |
| Flight 6 Complex--Slow Flight/Stalls | 44 | 1.5 | | | | | | | 1.5 | |
| Flight 7 Complex--Emergency Ops | 45 | 1.5 | | | | | | | 1.5 | |
| Ground 3 Airports, ATC, Airspace | 22 | | | | | | | | | 2.0 |
| Flight 8 Complex--Night Operations | 46 | 1.5 | | | | | 1.5 | | 1.5 | |
| Flight 9 Solo Practice | 47 | | 1.5 | | | | | | | |
| Flight 10 Performance Maneuvers | 48 | 1.5 | | | | | | | | |
| Ground 4 Federal Aviation Regulations | 23 | | | | | | | | | 3.0 |
| Flight 11 Eights-on-Pylons | 49 | 1.5 | | | | | | | | |
| Flight 12 Solo Practice | 50 | | 1.5 | | | | | | | |
| Flight 13 Solo Practice | 51 | | 1.5 | | | | | | | |
| Flight 14 Maneuvers Review | 52 | 1.5 | | | | | | | | |
| Ground 5 Performance, Wt. & Bal. | 24 | | | | | | | | | 3.0 |

| | | | | | | | | | | |
|---|---|---|---|---|---|---|---|---|---|---|
| Flight 15<br>Complex--Maneuvers Review | 53 | 1.5 | | | | | | | 1.5 | |
| Flight 16<br>Solo Practice | 54 | | 1.5 | | | | | | | |
| Ground Stage One<br>Knowledge Test | 25 | | | | | | | | | 1.0 |
| Flight 17<br>Stage One Check | 55 | 1.5 | | | | | | | 1.5 | |
| **STAGE ONE TOTALS** | | **17.5** | **7.5** | **0.0** | **0.0** | **1.0** | **1.5** | **0.0** | **8.5** | **15.0** |

| | | **Flight** | | | | | | | | **Ground** |
|---|---|---|---|---|---|---|---|---|---|---|
| LESSON | Page | Flight Training Dual | Flight Training Solo | Dual Cross-Country | Solo Cross-Country | Instrument | Night Dual | Night Solo | Complex Airplane Dual | Aeronautical Knowledge Training |
| **STAGE TWO** | | | | | | | | | | |
| Ground 6<br>Aeromedical/ADM | 26 | | | | | | | | | 2.5 |
| Ground 7<br>Aviation Weather | 27 | | | | | | | | | 3.0 |
| Flight 18<br>Maneuvers Review | 57 | 1.5 | | | | 0.5 | | | | |
| Ground 8<br>Aviation Weather Services | 28 | | | | | | | | | 3.0 |
| Flight 19<br>Dual Cross-Country | 58 | 4.0 | | 4.0 | | 1.5 | | | | |
| Flight 20<br>Night Flight--Local | 59 | 1.5 | | | | 0.5 | 1.5 | | | |
| Ground 9<br>Navigation: Charts, Publications, Flight Computers | 29 | | | | | | | | | 3.0 |
| Flight 21<br>Dual Night Cross-Country | 60 | 3.0 | | 3.0 | | 1.0 | 3.0 | | | |
| Flight 22<br>Solo Night--Local | 61 | | 1.5 | | | | | 1.5 | | |
| Ground 10<br>Navigation Systems | 30 | | | | | | | | | 2.5 |
| Flight 23<br>Solo Night Cross-Country | 62 | | 3.0 | | 3.0 | | | 3.0 | | |
| Flight 24<br>Solo Cross-Country | 63 | | 3.0 | | 3.0 | | | | | |
| Flight 25<br>Instrument Review | 64 | 1.5 | | | | 1.5 | | | | |
| Flight 26<br>Solo Cross-Country | 65 | | 5.0 | | 5.0 | | | | | |
| Flight 27<br>Solo Cross-Country | 66 | | 5.0 | | 5.0 | | | | | |
| Flight 28<br>Solo Night--Local | 67 | | 1.5 | | | | | 1.5 | | |
| Flight 29<br>Solo Cross-Country | 68 | | 5.0 | | 5.0 | | | | | |
| Flight 30<br>Solo Cross-Country | 69 | | 5.0 | | 5.0 | | | | | |
| Ground 11<br>Flight Operations | 31 | | | | | | | | | 2.0 |
| Flight 31<br>Maneuvers Review | 70 | 1.5 | | | | 0.5 | | | | |
| Flight 32<br>Solo Cross-Country | 71 | | 5.0 | | 5.0 | | | | | |
| Flight 33<br>Solo Cross-Country | 72 | | 7.0 | | 7.0 | | | | | |
| Ground Stage Two<br>Knowledge Test | 32 | | | | | | | | | 1.0 |
| Ground End-of-Course<br>Knowledge Test | 32 | | | | | | | | | 3.0 |
| Flight 34<br>Stage Two Check | 73 | 1.0 | | | | 0.5 | | | | |
| **STAGE TWO TOTALS** | | **14.0** | **41.0** | **7.0** | **38.0** | **6.0** | **4.5** | **6.0** | **0.0** | **20.0** |

| | | Flight | | | | | | | | Ground |
|---|---|---|---|---|---|---|---|---|---|---|
| LESSON | Page | Flight Training Dual | Flight Training Solo | Dual Cross-Country | Solo Cross-Country | Instrument | Night Dual | Night Solo | Complex Airplane Dual | Aeronautical Knowledge Training |
| **STAGE THREE** | | | | | | | | | | |
| Flight 35 Complex--Perf. Maneuvers | 75 | 1.5 | | | | | | | 1.5 | |
| Flight 36 Complex--Eights-on-Pylons | 76 | 1.5 | | | | 0.5 | | | 1.5 | |
| Flight 37 Solo Practice | 77 | | 1.5 | | | | | | | |
| Flight 38 Solo Practice | 78 | | 1.5 | | | | | | | |
| Flight 39 Maneuvers Review | 79 | 1.5 | | | | | | | | |
| Flight 40 Maneuvers Review | 80 | 1.5 | | | | | | | | |
| Flight 41 Solo Practice | 81 | | 1.5 | | | | | | | |
| Flight 42 Solo Practice | 82 | | 1.5 | | | | | | | |
| Flight 43 Complex--Maneuvers Review | 83 | 1.5 | | | | 0.5 | | | 1.5 | |
| Flight 44 Solo Practice | 84 | | 1.5 | | | | | | | |
| Flight 45 Maneuvers Review | 85 | 1.5 | | | | | | | | |
| Flight 46 Solo Practice | 86 | | 1.5 | | | | | | | |
| Flight 47 Maneuvers Review | 87 | 1.5 | | | | | | | | |
| Flight 48 Maneuvers Review | 88 | 1.5 | | | | 0.5 | | | | |
| Flight 49 Solo Practice | 89 | | 1.5 | | | | | | | |
| Flight 50 Solo Practice | 90 | | 1.5 | | | | | | | |
| Flight 51 Complex--Maneuvers Review | 91 | 1.5 | | | | | | | 1.5 | |
| Flight 52 Solo Practice | 92 | | 1.5 | | | | | | | |
| Flight 53 Maneuvers Review | 93 | 1.5 | | | | 0.5 | | | | |
| Flight 54 Dual Cross-Country | 94 | 3.0 | | 3.0 | | 1.0 | | | | |
| Flight 55 Solo Practice | 95 | | 1.5 | | | | | | | |
| Flight 56 Maneuvers Review | 96 | 1.0 | | | | | | | | |
| Flight 57 Solo Practice | 97 | | 1.5 | | | | | | | |
| Flight 58 Complex--Maneuvers Review | 98 | 1.0 | | | | | | | 1.0 | |
| Flight 59 Stage Three Check | 99 | 1.5 | | | | | | | 1.5 | |
| Flight 60 End-of-Course Check | 100 | 2.0 | | | | | | | 2.0 | |
| **STAGE THREE TOTALS** | | **23.5** | **16.5** | **3.0** | **0.0** | **3.0** | **0.0** | **0.0** | **10.5** | **0.0** |
| **COURSE TOTALS** | | **55.0** | **65.0** | **10.0**** | **38.0**** | **10.0** | **6.0** | **6.0** | **19.0** | **35.0** |
| | | **TOTAL TIMES*** | | | | | | | | |

Part 61 requires:

* 250 total logged hours
** 50 hours of total PIC cross country time

(Increase lesson times or repeat lessons as necessary to meet the minimum total time requirement and reach the appropriate level of pilot proficiency.)

NOTE: Any flight lesson in this training syllabus may be conducted in a complex airplane at the discretion of the flight instructor and/or chief flight instructor, as appropriate. We have indicated the minimum number of lessons that must be completed in a complex airplane. This syllabus does not require solo operations in a complex airplane, but solo lessons may also be conducted in a complex airplane at the discretion of the flight instructor and/or chief flight instructor, as appropriate.

# PART I: COMMERCIAL PILOT GROUND TRAINING SYLLABUS AIRPLANE SINGLE-ENGINE LAND

## GROUND TRAINING COURSE OBJECTIVES

The pilot will obtain the necessary aeronautical knowledge and meet the prerequisites specified in Appendix D to 14 CFR Part 141 (and Sec. 61.125) to successfully pass the commercial pilot (airplane) airman knowledge test.

## GROUND TRAINING COURSE COMPLETION STANDARDS

The pilot will demonstrate through stage knowledge tests and school records that (s)he meets the prerequisites specified in Appendix D to 14 CFR Part 141 (and Sec. 61.125) and has the aeronautical knowledge necessary to pass the commercial pilot (airplane) airman knowledge test.

| | | | | *Commercial Pilot Knowledge Test* | |
|---|---|---|---|---|---|
| Lesson | Topic | Min. Time in Hours | Reading Assignment No. of Pgs. | No. of O/L Pgs. | No. of Questions |
| | **Stage One** | | | | |
| 1 | Airplanes and Aerodynamics | 3.0 | 54 | 7 | 85 |
| 2 | Airplane Instruments, Engines, and Systems | 3.0 | 62 | 5 | 47 |
| 3 | Airports, Air Traffic Control, and Airspace | 2.0 | 72 | 9 | 69 |
| 4 | Federal Aviation Regulations | 3.0 | 120 | 11 | 136 |
| 5 | Airplane Performance and Weight and Balance | 3.0 | 30 | 10 | 65 |
| | Stage One Knowledge Test | 1.0 | | | |
| | **Stage Two** | | | | |
| 6 | Aeromedical Factors and Aeronautical Decision Making (ADM) | 2.5 | 34 | 4 | 47 |
| 7 | Aviation Weather | 3.0 | 147 | 5 | 96 |
| 8 | Aviation Weather Services | 3.0 | 146 | 6 | 49 |
| 9 | Navigation: Charts, Publications, Flight Computers | 3.0 | 51 | 6 | 63 |
| 10 | Navigation Systems | 2.5 | 34 | 5 | 39 |
| 11 | Flight Operations | 2.0 | 28 | 3 | 23 |
| | Stage Two Knowledge Test | 1.0 | | | |
| | End-of-Course Knowledge Test | 3.0 | | | |

## STAGE ONE

### Stage One Objective

To further develop the pilot's knowledge of airplanes and the aerodynamic principles of flight. The pilot will learn about the operation of various airplane systems, airport operations, radio communication procedures, air traffic control (ATC) radar services, and the National Airspace System (NAS). Additionally, the pilot will become familiar with pertinent Federal Aviation Regulations (FARs) and the accident-reporting requirements of the National Transportation Safety Board (NTSB). Finally, the pilot will review the use of performance charts and weight and balance computations.

### Stage One Completion Standards

Stage One will have been successfully completed when the pilot passes the Stage One knowledge test with a minimum passing grade of 80%.

| Lesson | Topic | Min. Time |
|---|---|---|
| 1 | Airplanes and Aerodynamics | 3.0 |
| 2 | Airplane Instruments, Engines, and Systems | 3.0 |
| 3 | Airports, Air Traffic Control, and Airspace | 2.0 |
| 4 | Federal Aviation Regulations | 3.0 |
| 5 | Airplane Performance and Weight and Balance | 3.0 |
| | Stage One Knowledge Test | 1.0 |

## GROUND LESSON 1: AIRPLANES AND AERODYNAMICS

### Objective

To further develop the pilot's knowledge of basic aerodynamics and the principles of flight.

### Text References

*Pilot Handbook*, Study Unit 1, "Airplanes and Aerodynamics" (54 pages)
*Commercial Pilot FAA Knowledge Test*, Study Unit 1, "Airplanes and Aerodynamics" (25 pages)

| *Pilot Handbook* Study Unit 1 Contents | Page(s) | *Commercial Pilot FAA Knowledge Test* Study Unit 1 Contents | No. of Questions |
|---|---|---|---|
| *1.1 Definitions* | *25-29* | *1.1 Flaps* | *2* |
| *1.2 The Airplane* | *30-34* | *1.2 Airplane Wings* | *7* |
| *1.3 Axes of Rotation* | *35* | *1.3 Stalls* | *14* |
| *1.4 Flight Controls and Control Surfaces* | *36-41* | *1.4 Spins* | *1* |
| *1.5 Forces Acting on the Airplane in Flight* | *41-47* | *1.5 Lift and Drag* | *25* |
| *1.6 Dynamics of the Airplane in Flight* | *47-51* | *1.6 Ground Effect* | *3* |
| *1.7 Ground Effect* | *51-52* | *1.7 Airplane Stability* | *5* |
| *1.8 How Airplanes Turn* | *53-54* | *1.8 Turns* | *8* |
| *1.9 Torque (Left-Turning Tendency)* | *54-58* | *1.9 Load Factor* | *16* |
| *1.10 Airplane Stability* | *58-64* | *1.10 Transonic and Supersonic Flight* | *4* |
| *1.11 Loads and Load Factors* | *65-70* | | |
| *1.12 Stalls and Spins* | *71-78* | | |

### Completion Standards

The lesson will have been successfully completed when the pilot answers the questions in Study Unit 1, "Airplanes and Aerodynamics," of *Commercial Pilot FAA Knowledge Test*, FAA Test Prep Online, and/or Online Ground School with a minimum passing grade of 80%.

---

| | Dates Studied | | | | | Date Completed |
|---|---|---|---|---|---|---|
| *Pilot Handbook* | ____ | ____ | ____ | ____ | ____ | ____ |
| *Commercial Pilot FAA Knowledge Test* | ____ | ____ | ____ | ____ | ____ | ____ |

Notes:

## GROUND LESSON 2: AIRPLANE INSTRUMENTS, ENGINES, AND SYSTEMS

### Objective

To further develop the pilot's knowledge of the airplane's instruments, engines, and systems.

### Text References

*Pilot Handbook*, Study Unit 2, "Airplane Instruments, Engines, and Systems" (62 pages)
*Commercial Pilot FAA Knowledge Test*, Study Unit 2, "Airplane Instruments, Engines, and Systems" (15 pages)

| *Pilot Handbook* Study Unit 2 Contents | Page(s) | *Commercial Pilot FAA Knowledge Test* Study Unit 2 Contents | No. of Questions |
|---|---|---|---|
| *2.1 Pitot-Static System* | *79-80* | *2.1 Magnetic Compass* | *1* |
| *2.2 Altimeter* | *81-84* | *2.2 Airspeed Indicator* | *9* |
| *2.3 Vertical Speed Indicator* | *84* | *2.3 Turn Coordinator/Turn-and-Slip Indicator* | *2* |
| *2.4 Airspeed Indicator* | *85-89* | *2.4 Fuel/Air Mixture* | *10* |
| *2.5 Gyroscopic Flight Instruments* | *89-91* | *2.5 Carburetor Heat* | *3* |
| *2.6 Turn Coordinator* | *91-93* | *2.6 Detonation and Preignition* | *4* |
| *2.7 Turn-and-Slip Indicator* | *93* | *2.7 Airplane Ignition Systems* | *5* |
| *2.8 Attitude Indicator* | *94* | *2.8 Engine Cooling* | *3* |
| *2.9 Heading Indicator* | *95-96* | *2.9 Airplane Propellers* | *10* |
| *2.10 Magnetic Compass* | *96-97* | | |
| *2.11 Compass Errors* | *98-101* | | |
| *2.12 Glass Cockpit Instrumentation.* | *102-104* | | |
| *2.13 Airplane Engines* | *104-106* | | |
| *2.14 How an Engine Operates* | *106-107* | | |
| *2.15 Ignition System* | *108-110* | | |
| *2.16 Induction System* | *110-119* | | |
| *2.17 Fuel System* | *119-123* | | |
| *2.18 Oil System* | *124-125* | | |
| *2.19 Cooling System* | *125-126* | | |
| *2.20 Propellers* | *126-131* | | |
| *2.21 Full Authority Digital Engine Control (FADEC).* | *131-132* | | |
| *2.22 Electrical System* | *132-136* | | |
| *2.23 Landing Gear System* | *136-138* | | |
| *2.24 Environmental System* | *138-139* | | |
| *2.25 Deice and Anti-Ice Systems* | *139-140* | | |

### Completion Standards

The lesson will have been successfully completed when the pilot answers the questions in Study Unit 2, "Airplane Instruments, Engines, and Systems," of *Commercial Pilot FAA Knowledge Test*, FAA Test Prep Online, and/or Online Ground School with a minimum passing grade of 80%.

---

| | Dates Studied | | | | | Date Completed |
|---|---|---|---|---|---|---|
| *Pilot Handbook* | ____ | ____ | ____ | ____ | ____ | ____ |
| *Commercial Pilot FAA Knowledge Test* | ____ | ____ | ____ | ____ | ____ | ____ |

Notes:

## GROUND LESSON 3: AIRPORTS, AIR TRAFFIC CONTROL, AND AIRSPACE

### Objective

To further develop the pilot's knowledge of airports, wake turbulence and collision avoidance, and the National Airspace System.

### Text References

*Pilot Handbook*, Study Unit 3, "Airports, Air Traffic Control, and Airspace" (72 pages)
*Commercial Pilot FAA Knowledge Test*, Study Unit 3, "Airports, Air Traffic Control, and Airspace" (26 pages)

| *Pilot Handbook* Study Unit 3 Contents | Page(s) | *Commercial Pilot FAA Knowledge Test* Study Unit 3 Contents | No. of Questions |
|---|---|---|---|
| 3.1 *Runway and Taxiway Markings* | *142-155* | 3.1 *Airspace* | *20* |
| 3.2 *Airport Lighting* | *156-160* | 3.2 *VHF/DF* | *1* |
| 3.3 *Visual Glideslope Indicators* | *161-164* | 3.3 *Airport Signs/Markings* | *28* |
| 3.4 *Wind and Landing Direction Indicators and Segmented Circles* | *164-166* | 3.4 *Collision Avoidance* | *3* |
| 3.5 *Airport Traffic Patterns* | *166-168* | 3.5 *Wake Turbulence* | *8* |
| 3.6 *Land and Hold Short Operations (LAHSO)* | *168-169* | 3.6 *Land and Hold Short Operations (LAHSO)* | *9* |
| 3.7 *Wake Turbulence* | *169-176* | | |
| 3.8 *Collision Avoidance* | *176-181* | | |
| 3.9 *Radio Communications and Phraseology* | *181-184* | | |
| 3.10 *Airports without an Operating Control Tower* | *184-187* | | |
| 3.11 *Automated Weather Reporting Systems* | *187-188* | | |
| 3.12 *Airports with an Operating Control Tower* | *189* | | |
| 3.13 *Automatic Terminal Information Service (ATIS)* | *189-190* | | |
| 3.14 *Ground Control* | *190* | | |
| 3.15 *Tower Control* | *191* | | |
| 3.16 *Approach Control and Departure Control (for VFR Aircraft)* | *191-192* | | |
| 3.17 *Clearance Delivery* | *192* | | |
| 3.18 *Emergencies* | *192* | | |
| 3.19 *Radio Failure Procedures* | *193-194* | | |
| 3.20 *Emergency Locator Transmitter (ELT)* | *194-195* | | |
| 3.21 *ATC Radar* | *196* | | |
| 3.22 *Transponder Operation* | *197-199* | | |
| 3.23 *Radar Services to VFR Aircraft* | *199-201* | | |
| 3.24 *General Dimensions of Airspace* | *201-202* | | |
| 3.25 *Controlled and Uncontrolled Airspace* | *202-204* | | |
| 3.26 *Class A Airspace* | *204* | | |
| 3.27 *Class B Airspace* | *204-205* | | |
| 3.28 *Class C Airspace* | *205* | | |
| 3.29 *Class D Airspace* | *206* | | |
| 3.30 *Class E Airspace* | *206-207* | | |
| 3.31 *Class G Airspace* | *207* | | |
| 3.32 *Special-Use Airspace* | *208-209* | | |
| 3.33 *Other Airspace Areas* | *209-211* | | |
| 3.34 *Special Flight Rules Areas* | *211-212* | | |

### Completion Standards

The lesson will have been successfully completed when the pilot answers the questions in Study Unit 3, "Airports, Air Traffic Control, and Airspace," of *Commercial Pilot FAA Knowledge Test*, FAA Test Prep Online, and/or Online Ground School with a minimum passing grade of 80%.

---

| | Dates Studied | | | | | Date Completed |
|---|---|---|---|---|---|---|
| *Pilot Handbook* | _____ | _____ | _____ | _____ | _____ | _____ |
| *Commercial Pilot FAA Knowledge Test* | _____ | _____ | _____ | _____ | _____ | _____ |

Notes:

## GROUND LESSON 4: FEDERAL AVIATION REGULATIONS

### Objective

To further develop the pilot's knowledge of pertinent federal regulations that relate to commercial pilots and the accident-reporting rules of the National Transportation Safety Board (NTSB).

### Text References

*FAR/AIM* (120 pages)
*Commercial Pilot FAA Knowledge Test*, Study Unit 4, "Federal Aviation Regulations" (42 pages)

| *FAR/AIM* Contents | Sections | *Commercial Pilot FAA Knowledge Test* Study Unit 4 Contents | No. of Questions |
|---|---|---|---|
| *Part 1 -- Definitions and Abbreviations* | *Entire Part* | *4.1 FAR Part 1* | *12* |
| *Part 61 -- Certification: Pilots, Flight Instructors, and Ground Instructors* | *Entire Part* | *4.2 FAR Part 23* | *1* |
| *Part 91 -- General Operating and Flight Rules* | *91.1-.417* | *4.3 FAR Part 61* | *27* |
| *Part 119 -- Certification: Air Carriers and Commercial Operators* | *119.1* | *4.4 FAR Part 91* | *81* |
| *NTSB Part 830 -- Notification and Reporting of Aircraft Accidents or Incidents and Overdue Aircraft, and Preservation of Aircraft Wreckage, Mail, Cargo, and Records* | *Entire Part* | *4.5 FAR Part 119* | *2* |
| | | *4.6 NTSB Part 830* | *11* |
| | | *4.7 Near Midair Collision Reporting* | *2* |

### Completion Standards

The lesson will have been successfully completed when the pilot answers the questions in Study Unit 4, "Federal Aviation Regulations," of *Commercial Pilot FAA Knowledge Test*, FAA Test Prep Online, and/or Online Ground School with a minimum passing grade of 80%.

---

| | Dates Studied | | | | | Date Completed |
|---|---|---|---|---|---|---|
| *FAR/AIM* | ____ | ____ | ____ | ____ | ____ | ____ |
| *Commercial Pilot FAA Knowledge Test* | ____ | ____ | ____ | ____ | ____ | ____ |

Notes:

## GROUND LESSON 5: AIRPLANE PERFORMANCE AND WEIGHT AND BALANCE

### Objective

To further develop the pilot's ability to determine airplane performance, including weight and balance. Additionally, the pilot will learn the significance and effects of exceeding the airplane's performance limitations.

### Text References

*Pilot Handbook*, Study Unit 5, "Airplane Performance and Weight and Balance" (30 pages)
*Commercial Pilot FAA Knowledge Test*, Study Unit 5, "Airplane Performance and Weight and Balance" (46 pages)

| *Pilot Handbook* Study Unit 5 Contents | Page(s) |
|---|---|
| 5.1 Determinants of Airplane Performance | 297-299 |
| 5.2 Standard Atmosphere | 299-300 |
| 5.3 Pressure Altitude | 300 |
| 5.4 Density Altitude | 301-303 |
| 5.5 Takeoff Performance | 304-306 |
| 5.6 Climb Performance | 306-309 |
| 5.7 Cruise and Range Performance | 309-310 |
| 5.8 Glide Performance | 310-311 |
| 5.9 Crosswind Performance | 311-312 |
| 5.10 Landing Performance | 312-315 |
| 5.11 Weight and Balance Overview | 315-317 |
| 5.12 Weight and Balance Terms | 317-318 |
| 5.13 Basic Principles of Weight and Balance | 318-319 |
| 5.14 Methods of Determining Weight and Balance | 319 |
| 5.15 Center of Gravity Calculations | 320 |
| 5.16 Center of Gravity Charts | 320-322 |
| 5.17 Center of Gravity Tables | 323-325 |
| 5.18 Weight Change and Weight Shift Computations | 326 |

| *Commercial Pilot FAA Knowledge Test* Study Unit 5 Contents | No. of Questions |
|---|---|
| 5.1 Density Altitude | 4 |
| 5.2 Density Altitude Computations | 5 |
| 5.3 Takeoff Distance | 5 |
| 5.4 Time, Fuel, and Distance to Climb | 10 |
| 5.5 Maximum Rate of Climb | 2 |
| 5.6 Cruise and Range Performance | 15 |
| 5.7 Crosswind/Headwind Component | 4 |
| 5.8 Landing Distance | 4 |
| 5.9 Weight and Balance | 4 |
| 5.10 Weight and Moment Computations | 8 |
| 5.11 Weight Change and Weight Shift Computations | 4 |

### Completion Standards

The lesson will have been successfully completed when the pilot answers the questions in Study Unit 5, "Airplane Performance and Weight and Balance," of *Commercial Pilot FAA Knowledge Test*, FAA Test Prep Online, and/or Online Ground School with a minimum passing grade of 80%.

---

| | Dates Studied | | | | | Date Completed |
|---|---|---|---|---|---|---|
| *Pilot Handbook* | _____ | _____ | _____ | _____ | _____ | _____ |
| *Commercial Pilot FAA Knowledge Test* | _____ | _____ | _____ | _____ | _____ | _____ |

Notes:

## STAGE ONE KNOWLEDGE TEST

### Objective

To evaluate the pilot's understanding of the material presented during Ground Lesson 1 through Ground Lesson 5. The Stage One knowledge test consists of 25 questions.

### Content

| Lesson | |
|---|---|
| 1 | Airplanes and Aerodynamics |
| 2 | Airplane Instruments, Engines, and Systems |
| 3 | Airports, Air Traffic Control, and Airspace |
| 4 | Federal Aviation Regulations |
| 5 | Airplane Performance and Weight and Balance |

### Completion Standards

The lesson will have been successfully completed when the pilot has completed the Stage One knowledge test with a minimum passing grade of 80%.

## STAGE TWO

### Stage Two Objective

To further develop the pilot's knowledge of aeromedical factors and the aeronautical decision-making process related to all flights. The pilot will learn how weather affects flying, including how to recognize and avoid critical weather situations such as wind shear. The pilot will be able to interpret aviation reports, forecasts, and charts. Additionally, the pilot will review the use of navigation charts, plotters, flight computers, and flight publications for cross-county flight planning. Finally, the pilot will review various navigation systems and maneuvers, procedures, and emergency operations.

### Stage Two Completion Standards

Stage Two will have been successfully completed when the pilot passes the Stage Two knowledge test with a minimum passing grade of 80%.

| Lesson | Topic | Min. Time |
|---|---|---|
| 6 | Aeromedical Factors and Aeronautical Decision Making (ADM) | 2.5 |
| 7 | Aviation Weather | 3.0 |
| 8 | Aviation Weather Services | 3.0 |
| 9 | Navigation: Charts, Publications, Flight Computers | 3.0 |
| 10 | Navigation Systems | 2.5 |
| 11 | Flight Operations | 2.0 |
| | Stage Two Knowledge Test | 1.0 |
| | End-of-Course Knowledge Test | 3.0 |

## GROUND LESSON 6: AEROMEDICAL FACTORS AND AERONAUTICAL DECISION MAKING (ADM)

### Objective

To further develop the pilot's knowledge of the medical factors related to flight and to the aeronautical decision-making (ADM) process.

### Text References

*Pilot Handbook*, Study Unit 6, "Aeromedical Factors and Aeronautical Decision Making (ADM)" (34 pages)

*Commercial Pilot FAA Knowledge Test*, Study Unit 6, "Aeromedical Factors and Aeronautical Decision Making (ADM)" (13 pages)

| *Pilot Handbook* Study Unit 6 Contents | Page(s) | *Commercial Pilot FAA Knowledge Test* Study Unit 6 Contents | No. of Questions |
|---|---|---|---|
| *6.1 Fitness for Flight* | *327-331* | *6.1 Hypoxia and Alcohol* | *12* |
| *6.2 Hypoxia* | *331-333* | *6.2 Hyperventilation* | *4* |
| *6.3 Dehydration* | *333* | *6.3 Spatial Disorientation* | *4* |
| *6.4 Hyperventilation* | *333-334* | *6.4 Pilot Vision* | *4* |
| *6.5 Carbon Monoxide Poisoning* | *334-335* | *6.5 Aeronautical Decision Making (ADM)* | *23* |
| *6.6 Decompression Sickness after Scuba Diving* | *336* | | |
| *6.7 Motion Sickness* | *336* | | |
| *6.8 Sinus and Ear Block* | *337* | | |
| *6.9 Spatial Disorientation* | *338* | | |
| *6.10 Illusions in Flight* | *338-340* | | |
| *6.11 Vision* | *340-343* | | |
| *6.12 Aeronautical Decision Making (ADM)* | *343-349* | | |
| *6.13 Weather-Related Decision Making* | *350-352* | | |
| *6.14 Stress and Flying* | *352-355* | | |
| *6.15 Identifying the Enemy* | *355-357* | | |
| *6.16 Single-Pilot Resource Management (SRM)* | *357-360* | | |

### Completion Standards

The lesson will have been successfully completed when the pilot answers the questions in Study Unit 6, "Aeromedical Factors and Aeronautical Decision Making (ADM)," of *Commercial Pilot FAA Knowledge Test*, FAA Test Prep Online, and/or Online Ground School with a minimum passing grade of 80%.

---

| | Dates Studied | | | | | Date Completed |
|---|---|---|---|---|---|---|
| *Pilot Handbook* | ____ | ____ | ____ | ____ | ____ | ____ |
| *Commercial Pilot FAA Knowledge Test* | ____ | ____ | ____ | ____ | ____ | ____ |

Notes:

## GROUND LESSON 7: AVIATION WEATHER

### Objective

To further develop the pilot's knowledge of weather, as related to flight operations. Additionally, the pilot will be able to recognize critical weather situations and wind shear recognition and avoidance.

### Text References

*Aviation Weather and Weather Services*, Study Units 1-13 (Parts I and II) (147 pages)
*Commercial Pilot FAA Knowledge Test*, Study Unit 7, "Aviation Weather" (24 pages)

| *Aviation Weather and Weather Services* Reading Assignment | Pages | *Commercial Pilot FAA Knowledge Test* Study Unit 7 Contents | No. of Questions |
|---|---|---|---|
| *The Earth's Atmosphere* | 3-6 | *7.1 Causes of Weather* | 8 |
| *Temperature* | 7-13 | *7.2 High/Low Pressure Areas* | 7 |
| *Atmospheric Pressure and Altimetry* | 15-28 | *7.3 Jet Stream* | 7 |
| *Wind* | 29-44 | *7.4 Temperature* | 5 |
| *Moisture, Cloud Formation, and Precipitation* | 45-54 | *7.5 Clouds* | 7 |
| *Stable and Unstable Air* | 55-62 | *7.6 Fog* | 8 |
| *Clouds* | 63-72 | *7.7 Stability* | 17 |
| *Air Masses and Fronts* | 73-90 | *7.8 Thunderstorms and Icing* | 21 |
| *Turbulence* | 91-102 | *7.9 Turbulence* | 8 |
| *Icing* | 103-112 | *7.10 Wind Shear* | 8 |
| *Thunderstorms* | 113-130 | | |
| *Common IFR Procedures* | 131-140 | | |
| *High Altitude Weather* | 143-152 | | |

### Completion Standards

The lesson will have been successfully completed when the pilot answers the questions in Study Unit 7, "Aviation Weather," of *Commercial Pilot FAA Knowledge Test*, FAA Test Prep Online, and/or Online Ground School with a minimum passing grade of 80%.

---

| | Dates Studied | | | | | Date Completed |
|---|---|---|---|---|---|---|
| *Aviation Weather and Weather Services* | _____ | _____ | _____ | _____ | _____ | _____ |
| *Commercial Pilot FAA Knowledge Test* | _____ | _____ | _____ | _____ | _____ | _____ |

Notes:

## GROUND LESSON 8: AVIATION WEATHER SERVICES

### Objective

To further develop the pilot's ability to interpret and use weather charts, reports, forecasts, and broadcasts.

### Text References

*Aviation Weather and Weather Services*, Part III, "Aviation Weather Services" (146 pages)
*Commercial Pilot FAA Knowledge Test*, Study Unit 8, "Aviation Weather Services" (17 pages)

| *Aviation Weather and Weather Services* Part III: Reading Assignment | Pages | *Commercial Pilot FAA Knowledge Test* Study Unit 8 Contents | No. of Questions |
|---|---|---|---|
| *The Aviation Weather Service Program* | *211-220* | *8.1 Sources of Weather Information* | *6* |
| *Aviation Routine Weather Reports (METAR)* | *225-246* | *8.2 Aviation Routine Weather Report (METAR)* | *7* |
| *Pilot Weather Reports (PIREP)* | *247-252* | *8.3 Radar Weather Report (SD/ROB)* | *1* |
| *Radar Weather Reports (SD/ROB)* | *253-256* | *8.4 Surface Analysis Chart* | *5* |
| *Surface Analysis Charts* | *277-282* | *8.5 Constant Pressure Charts* | *3* |
| *Constant Pressure Charts* | *283-292* | *8.6 Terminal Aerodrome Forecast (TAF)* | *5* |
| *Lifted Index Analysis Charts* | *297-300* | *8.7 Aviation Area Forecast (FA)* | *4* |
| *Weather Depiction Charts* | *301-304* | *8.8 In-Flight Weather Advisories* | *7* |
| *Radar Summary Charts* | *311-318* | *8.9 Low-Level and High-Level Prognostic Charts* | *4* |
| *Significant Meteorological Information (SIGMET)* | *327-340* | *8.10 Other Charts and Forecasts* | *7* |
| *Airmen's Meteorological Information (AIRMET)* | *341-348* | | |
| *Center Weather Advisory (CWA)* | *349-352* | | |
| *Area Forecasts (FA)* | *379-392* | | |
| *Terminal Aerodrome Forecast (TAF)* | *393-404* | | |
| *Wind and Temperature Aloft Forecasts (FB)* | *407-412* | | |
| *Short-Range Surface Prognostic (PROG) Charts* | *413-416* | | |
| *Low-Level Significant Weather (SIGWX) Charts* | *417-420* | | |
| *High-Level Significant Weather (SIGWX) Charts* | *429-434* | | |

### Completion Standards

The lesson will have been successfully completed when the pilot answers the questions in Study Unit 8, "Aviation Weather Services," of *Commercial Pilot FAA Knowledge Test*, FAA Test Prep Online, and/or Online Ground School with a minimum passing grade of 80%.

---

| | Dates Studied | | | | | Date Completed |
|---|---|---|---|---|---|---|
| *Aviation Weather and Weather Services* | _____ | _____ | _____ | _____ | _____ | _____ |
| *Commercial Pilot FAA Knowledge Test* | _____ | _____ | _____ | _____ | _____ | _____ |

Notes:

## GROUND LESSON 9: NAVIGATION: CHARTS, PUBLICATIONS, FLIGHT COMPUTERS

### Objective

To further develop the pilot's knowledge of, and the ability to use, navigation charts, publications, and a flight computer in planning a VFR cross-country flight.

### Text References

*Pilot Handbook*, Study Unit 9, "Navigation: Charts, Publications, Flight Computers" (51 pages)
*Commercial Pilot FAA Knowledge Test*, Study Unit 9, "Navigation: Charts, Publications, Flight Computers" (24 pages)

| *Pilot Handbook* Study Unit 9 Contents | Page(s) | *Commercial Pilot FAA Knowledge Test* Study Unit 9 Contents | No. of Questions |
|---|---|---|---|
| *9.1 VFR Navigation Charts* | *429-431* | *9.1 Sectional Charts* | *18* |
| *9.2 Longitude and Latitude* | *431-434* | *9.2 Instrument Approach Charts* | *1* |
| *9.3 Sectional Chart Symbology* | *435-442* | *9.3 Fuel Consumption* | *8* |
| *9.4 FAA Advisory Circulars (AC)* | *443* | *9.4 Time, Distance, and Fuel to Station* | *26* |
| *9.5 Aeronautical Information Manual (AIM)* | *443-444* | *9.5 Wind Direction and Speed* | *4* |
| *9.6 Airport/Facility Directory (A/FD)* | *445-446* | *9.6 Time, Compass Heading, Etc., on Climbs and En Route* | *3* |
| *9.7 Notice to Airmen (NOTAM) System* | *447-450* | *9.7 Time, Compass Heading, Etc., on Descents* | *3* |
| *9.8 Flight Computers* | *451* | | |
| *9.9 The Gleim Flight Computer* | *451-452* | | |
| *9.10 The Calculator Side of the Flight Computer* | *452-453* | | |
| *9.11 Conversion of Nautical Miles to Statute Miles and Vice Versa* | *453-454* | | |
| *9.12 Speed, Distance, and Time Computations* | *454-458* | | |
| *9.13 Fuel Computations* | *459-461* | | |
| *9.14 True Airspeed and Density Altitude* | *462-463* | | |
| *9.15 Corrected (Approximately True) Altitude* | *464-465* | | |
| *9.16 Off-Course Correction* | *465-466* | | |
| *9.17 Radius of Action* | *466* | | |
| *9.18 Other Conversions* | *466-468* | | |
| *9.19 Temperature Conversions* | *468* | | |
| *9.20 The Wind Side of the Gleim Flight Computer* | *468-469* | | |
| *9.21 Determining Magnetic Heading and Groundspeed* | *469-471* | | |
| *9.22 Determining Wind Direction and Speed* | *472* | | |
| *9.23 Determining Altitude for Most Favorable Winds* | *472-473* | | |
| *9.24 Alternative: E6B Computer Approach to Magnetic Heading* | *473-474* | | |
| *9.25 Information Side of Sliding Card (Gleim E6B)* | *474-475* | | |
| *9.26 Electronic Flight Computers* | *476* | | |
| *9.27 ASA CX-2* | *477-478* | | |
| *9.28 Sporty's E6B* | *478-479* | | |

### Completion Standards

The lesson will have been successfully completed when the pilot answers the questions in Study Unit 9, "Navigation: Charts, Publications, Flight Computers," of *Commercial Pilot FAA Knowledge Test*, FAA Test Prep Online, and/or Online Ground School with a minimum passing grade of 80%.

| | Dates Studied | | | | | Date Completed |
|---|---|---|---|---|---|---|
| *Pilot Handbook* | ____ | ____ | ____ | ____ | ____ | ____ |
| *Commercial Pilot FAA Knowledge Test* | ____ | ____ | ____ | ____ | ____ | ____ |

Notes:

## GROUND LESSON 10: NAVIGATION SYSTEMS

### Objective

To further develop the pilot's knowledge of various navigation systems.

### Text References

*Pilot Handbook*, Study Unit 10, "Navigation Systems" (34 pages)
*Commercial Pilot FAA Knowledge Test*, Study Unit 10, "Navigation Systems" (18 pages)

| *Pilot Handbook* Study Unit 10 Contents | Page(s) | *Commercial Pilot FAA Knowledge Test* Study Unit 10 Contents | No. of Questions |
|---|---|---|---|
| *10.1 Basic Radio Principles* | *481-485* | *10.1 Automatic Direction Finder (ADF)* | *15* |
| *10.2 VHF Omnidirectional Range (VOR)* | *486-498* | *10.2 VOR Use and Receiver Checks* | *9* |
| *10.3 Distance-Measuring Equipment (DME)* | *498-499* | *10.3 Radio Magnetic Indicator (RMI)* | *5* |
| *10.4 Automatic Direction Finder (ADF)* | *500-506* | *10.4 Horizontal Situation Indicator (HSI)* | *5* |
| *10.5 Area Navigation (RNAV)* | *506* | *10.5 Global Positioning System* | *5* |
| *10.6 VORTAC-Based RNAV* | *506-507* | | |
| *10.7 Global Positioning System (GPS)* | *508-514* | | |

### Completion Standards

The lesson will have been successfully completed when the pilot answers the questions in Study Unit 10, "Navigation Systems," of *Commercial Pilot FAA Knowledge Test*, FAA Test Prep Online, and/or Online Ground School with a minimum passing grade of 80%.

---

| | Dates Studied | | | | | Date Completed |
|---|---|---|---|---|---|---|
| *Pilot Handbook* | ____ | ____ | ____ | ____ | ____ | ____ |
| *Commercial Pilot FAA Knowledge Test* | ____ | ____ | ____ | ____ | ____ | ____ |

Notes:

## GROUND LESSON 11: FLIGHT OPERATIONS

### Objective

To further develop the pilot's knowledge of maneuvers, procedures, and emergency operations.

### Text References

*Commercial Pilot Flight Maneuvers and Practical Test Prep* (28 pages)
*Commercial Pilot FAA Knowledge Test*, Study Unit 11, "Flight Operations" (8 pages)

| *Commercial Pilot Flight Maneuvers* Reading Assignment | Pages | *Commercial Pilot FAA Knowledge Test* Study Unit 11 Contents | No. of Questions |
|---|---|---|---|
| *Task II.D., Taxiing* | *132-138* | *11.1 Flight Fundamentals* | *1* |
| *Task IV.B., Normal and Crosswind Approach and Landing* | *179-196* | *11.2 Taxiing* | *3* |
| *Task VII.C., Diversion* | *277-279* | *11.3 Landings* | *3* |
| | | *11.4 Emergencies* | *2* |
| | | *11.5 Anti-Collision Light System* | *1* |
| | | *11.6 Cold Weather Operation* | *3* |
| | | *11.7 Turbulence* | *3* |
| | | *11.8 Night Flying Operations* | *7* |

### Completion Standards

The lesson will have been successfully completed when the pilot answers the questions in Study Unit 11, "Flight Operations," of *Commercial Pilot FAA Knowledge Test*, FAA Test Prep Online, and/or Online Ground School with a minimum passing grade of 80%.

---

| | Dates Studied | | | | | Date Completed |
|---|---|---|---|---|---|---|
| *Commercial Pilot Flight Maneuvers* | ____ | ____ | ____ | ____ | ____ | ____ |
| *Commercial Pilot FAA Knowledge Test* | ____ | ____ | ____ | ____ | ____ | ____ |

Notes:

## STAGE TWO KNOWLEDGE TEST

### Objective

To evaluate the pilot's understanding of the material presented during Ground Lesson 6 through Ground Lesson 11. The Stage Two knowledge test consists of 25 questions.

### Content

| <u>Lesson</u> | (Lessons 1-5 were Stage One.) |
|---|---|
| 6 | Aeromedical Factors and Aeronautical Decision Making (ADM) |
| 7 | Aviation Weather |
| 8 | Aviation Weather Services |
| 9 | Navigation: Charts, Publications, Flight Computers |
| 10 | Navigation Systems |
| 11 | Flight Operations |

### Completion Standards

This lesson will have been successfully completed when the pilot has completed the Stage Two knowledge test with a minimum passing grade of 80%.

## END-OF-COURSE KNOWLEDGE TEST

### Objective

To evaluate the pilot's comprehension of the material covered in the ground training course and to determine the pilot's readiness to take the FAA commercial pilot (airplane) knowledge test. The end-of-course knowledge test consists of 100 questions.

### Content

Practice Commercial Pilot Knowledge Test

### Completion Standards

The lesson will have been successfully completed when the pilot has completed the practice commercial pilot (airplane) knowledge test with a minimum passing grade of 80%.

# PART I: COMMERCIAL PILOT FLIGHT TRAINING SYLLABUS AIRPLANE SINGLE-ENGINE LAND

## FLIGHT TRAINING COURSE OBJECTIVES

The pilot will obtain the aeronautical knowledge and experience and demonstrate the flight proficiency necessary to meet the requirements for a commercial pilot certificate with an airplane category rating and single-engine land class rating.

## FLIGHT TRAINING COURSE COMPLETION STANDARDS

The pilot will demonstrate through the stage checks and school records that (s)he has the necessary flight proficiency and aeronautical experience to obtain a commercial pilot certificate with an airplane category rating and single-engine class rating.

| Lesson | Topic | Number of Pages in "New Item" Reading Assignment |
|---|---|---|
| | **Stage One** | |
| 1 | Traffic Pattern Operations | 129 |
| 2 | Slow Flight, Stalls, and Emergency Operations | 60 |
| 3 | Solo Practice | * |
| 4 | Maneuvers Review | * |
| 5 | Introduction to Complex Airplanes | 97 |
| 6 | Complex Airplane--Slow Flight and Stalls | 53 |
| 7 | Complex Airplane--Emergency Operations | 30 |
| 8 | Complex Airplane--Night Operations | 112 |
| 9 | Solo Practice | * |
| 10 | Introduction to Performance Maneuvers | 22 |
| 11 | Introduction to Eights-on-Pylons | 6 |
| 12 | Solo Practice | * |
| 13 | Solo Practice | * |
| 14 | Maneuvers Review | * |
| 15 | Complex Airplane--Maneuvers Review | * |
| 16 | Solo Practice | * |
| 17 | Stage One Check--Complex Airplane | * |
| | | 509 |
| | **Stage Two** | |
| 18 | Maneuvers Review | * |
| 19 | Dual Cross-Country | 81 |
| 20 | Night Flight--Local | * |
| 21 | Dual Night Cross-Country | 73 |
| 22 | Solo Night--Local | * |
| 23 | Solo Night Cross-Country | * |
| 24 | Solo Cross-Country | * |
| 25 | Instrument Review | * |
| 26 | Solo Cross-Country | * |
| 27 | Solo Cross-Country | * |
| 28 | Solo Night--Local | * |
| 29 | Solo Cross-Country | * |
| 30 | Solo Cross-Country | * |
| 31 | Maneuvers Review | * |
| 32 | Solo Cross-Country | * |
| 33 | Solo Cross-Country | * |
| 34 | Stage Two Check | * |
| | | 154 |
| | **Stage Three** | |
| 35 | Complex Airplane--Performance Maneuvers | 22 |
| 36 | Complex Airplane--Eights-on-Pylons | 6 |
| 37 | Solo Practice | * |
| 38 | Solo Practice | * |
| 39 | Maneuvers Review | * |
| 40 | Maneuvers Review | * |
| 41 | Solo Practice | * |
| 42 | Solo Practice | * |
| 43 | Complex Airplane--Maneuvers Review | * |
| 44 | Solo Practice | * |
| 45 | Maneuvers Review | * |
| 46 | Solo Practice | * |
| 47 | Maneuvers Review | * |
| 48 | Maneuvers Review | * |
| 49 | Solo Practice | * |
| 50 | Solo Practice | * |
| 51 | Complex Airplane--Maneuvers Review | * |
| 52 | Solo Practice | * |
| 53 | Maneuvers Review | * |
| 54 | Dual Cross-Country | * |
| 55 | Solo Practice | * |
| 56 | Maneuvers Review | * |
| 57 | Solo Practice | * |
| 58 | Complex Airplane--Maneuvers Review | * |
| 59 | Stage Three Check--Complex Airplane | * |
| 60 | End-of-Course Check--Complex Airplane | * |
| | | 28 |

*Reading assignment consists of review only, no new materials.

The following is a brief description of the parts of each flight lesson in this syllabus:

**Objective:** We open each lesson with an objective, usually a sentence or two, to help you gain perspective and understand the goal for that particular lesson.

**Text References:** This section tells you which reference books you will need to study or refer to while mastering the tasks within the lesson. Abbreviations are given to facilitate the cross-referencing process.

**Content:** Each lesson contains a list of the tasks required to be completed before moving to the next lesson. A task may be listed as a "review item" (a task that was covered in a previous lesson) or as a "new item" (a task that is introduced to you for the first time in this course). Each task is preceded by three blank "checkoff" boxes, which may be used by your instructor to keep track of your progress and to indicate that each task was completed.

There are three boxes because it may take more than one flight to complete the lesson. Your instructor may mark the box(es) next to each task in one of the following methods (or any other method desired):

| | | |
|---|---|---|
| ✓ - task completed to lesson completion standards | D - demonstrated by instructor *<br>A - accomplished by you<br>S - safe/satisfactory<br>P - meets PTS standards | 1 - above lesson standard<br>2 - meets lesson standard<br>3 - below lesson standard |

Most tasks are followed by book and page references that tell you where to find the information you need to study to accomplish the task successfully.

The last task in each flight lesson is labeled "Additional items at CFI's discretion" and is followed by several blank lines. This area can be used to record any extra items that your instructor feels are appropriate to the lesson, taking into account such variables as weather, local operational considerations, and your progress as a student.

NOTE: Instructors are reminded not to limit themselves to the blank lines provided -- use as much of the page as you need.

**Completion Standards:** Based on these standards, your instructor determines how well you have met the objective of the lesson in terms of knowledge and skill.

**Instructor's Comments and Lesson Assignment:** Space is provided for your instructor's critique of the lesson, which you can refer to later. Your instructor may also write any specific assignment for the next lesson.

*System suggested by the U.S. Air Force flying clubs

## Reading Assignments for Flight Lessons

You are expected to be prepared for each flight lesson. Our reading assignments will help you understand what is going to happen and how and why you need to do everything BEFORE you go to the airport.

Each flight lesson in this book contains

- Objective
- Text references
- Content
    - ✓ Review items
    - ✓ New items
- Completion standards
- Comments, assignment, and notes

Next to each item in the Content section, we have provided the page number(s) to read in *Commercial Pilot Flight Maneuvers and Practical Test Prep* (FM) and/or *Pilot Handbook* (PH), and the section to read, if appropriate, in your airplane's Pilot's Operating Handbook (POH).

For the new items, you should read the material and attempt to understand the basic concepts. Try to anticipate and visualize the concepts and flight maneuvers. With this basic knowledge, your instructor can expand on the specific and finer points, especially when explaining how a task is done in your specific airplane.

After your flight lesson, task items are fresh in your mind; they will make sense, and you should be able to understand and learn more.

Study review items so you can explain them to your instructor and, at the completion of this syllabus, to your examiner.

After you study, relax and plan a time to begin reading to prepare for the next flight lesson.

## STAGE ONE

### Stage One Objective

The pilot will obtain the knowledge and skill required to safely operate a complex airplane and receive a complex airplane logbook endorsement (if applicable). Additionally, the pilot will be introduced to commercial flight maneuvers.

### Stage One Completion Standards

The stage will be completed when the pilot demonstrates proficiency in the operation of a complex airplane and receive a complex airplane logbook endorsement (if applicable).

| Lesson | Topic |
|---|---|
| 1 | Traffic Pattern Operations |
| 2 | Slow Flight, Stalls, and Emergency Operations |
| 3 | Solo Practice |
| 4 | Maneuvers Review |
| 5 | Introduction to Complex Airplanes |
| 6 | Complex Airplane--Slow Flight and Stalls |
| 7 | Complex Airplane--Emergency Operations |
| 8 | Complex Airplane--Night Operations |
| 9 | Solo Practice |
| 10 | Introduction to Performance Maneuvers |
| 11 | Introduction to Eights-on-Pylons |
| 12 | Solo Practice |
| 13 | Solo Practice |
| 14 | Maneuvers Review |
| 15 | Complex Airplane--Maneuvers Review |
| 16 | Solo Practice |
| 17 | Stage One Check--Complex Airplane |

## FLIGHT LESSON 1: TRAFFIC PATTERN OPERATIONS

### Objective

To introduce the pilot to the listed tasks with emphasis on traffic pattern operations.

### Text References

*Commercial Pilot Flight Maneuvers and Practical Test Prep* (FM)
Pilot's Operating Handbook (POH)

### Content

1. Preflight briefing
2. New items
   - ☐☐☐ Certificates and documents - FM 68-71
   - ☐☐☐ Airworthiness requirements - FM 72-76
   - ☐☐☐ Operation of systems - FM 106-110; POH-1, 7, 8, 9
   - ☐☐☐ Aeromedical factors - FM 111-116
   - ☐☐☐ Preflight inspection - FM 118-122; POH-4
   - ☐☐☐ Cockpit management - FM 123-126
   - ☐☐☐ Engine starting - FM 127-131; POH-4
   - ☐☐☐ Taxiing - FM 132-138
   - ☐☐☐ Runway incursion avoidance - FM 139-143
   - ☐☐☐ Before-takeoff check - FM 144-147; POH-4
   - ☐☐☐ Radio communications and ATC light signals - FM 150-152
   - ☐☐☐ Traffic patterns - FM 153-159
     - ☐☐☐ Runway incursion and collision avoidance - FM 153-154
   - ☐☐☐ Airport, runway, and taxiway signs, markings, and lighting - FM 160-167
   - ☐☐☐ Normal and crosswind takeoff and climb - FM 170-178; POH-4
   - ☐☐☐ Normal and crosswind approach and landing - FM 179-196; POH-4
   - ☐☐☐ Short-field takeoff and maximum performance climb - FM 208-214; POH-4
   - ☐☐☐ Short-field approach and landing - FM 215-221; POH-4
   - ☐☐☐ Power-off 180° accuracy approach and landing - FM 222-230
   - ☐☐☐ Emergency approach and landing (simulated) - FM 314-320; POH-3
   - ☐☐☐ After landing, parking, and securing - FM 336-339; POH-4
   - ☐☐☐ Additional items at CFI's discretion ____________________

     ____________________
3. Postflight critique and preview of next lesson

### Completion Standards

The lesson will have been successfully completed when the pilot demonstrates at least a private pilot skill level in the assigned tasks, with the exception of a power-off 180° approach and landing. The pilot will demonstrate understanding of the techniques required to perform power-off accuracy approaches and landings.

**Instructor's comments:** ____________________

____________________

**Lesson assignment:** ____________________

____________________

**Notes:** ____________________

____________________

____________________

____________________

____________________

____________________

____________________

## FLIGHT LESSON 2: SLOW FLIGHT, STALLS, AND EMERGENCY OPERATIONS

### Objective

To introduce the pilot to slow flight, stalls, and emergency operations.

### Text References

*Commercial Pilot Flight Maneuvers and Practical Test Prep* (FM)
*Pilot Handbook* (PH)
Pilot's Operating Handbook (POH)

### Content

1. Flight Lesson 1 complete? Yes ____ Copy of lesson placed in pilot's folder? Yes ____
2. Preflight briefing
3. Review items
   - ☐☐☐ Radio communications and ATC light signals - FM 150-152
   - ☐☐☐ Traffic patterns - FM 153-159
   - ☐☐☐ Runway incursion avoidance - FM 139-143
   - ☐☐☐ Emergency approach and landing (simulated) - FM 314-320; POH-3

   ______________________________
4. New items
   - ☐☐☐ Aeronautical decision making (ADM) - PH 343-349
   - ☐☐☐ Soft-field takeoff and climb - FM 197-202; POH-4
   - ☐☐☐ Soft-field approach and landing - FM 203-207; POH-4
   - ☐☐☐ Go-around/rejected landing - FM 231-235; POH-4
   - ☐☐☐ Maneuvering during slow flight - FM 284-288
   - ☐☐☐ Power-off stalls - FM 289-294; PH 71-75
   - ☐☐☐ Power-on stalls - FM 295-300; PH 71-75
   - ☐☐☐ Accelerated stalls - FM 301-305; PH 71-75
   - ☐☐☐ Spin awareness - FM 306-308; POH-2
   - ☐☐☐ Emergency descent - FM 311-313; POH-3
   - ☐☐☐ Systems and equipment malfunctions - FM 321-322; POH-3, 9
   - ☐☐☐ Emergency equipment and survival gear - FM 323-324; POH-7, 9
   - ☐☐☐ Straight-and-level (IR)* - CFI
   - ☐☐☐ Constant airspeed climbs and descents (IR)* - CFI
   - ☐☐☐ Turns to a heading (IR)* - CFI
   - ☐☐☐ Recovery from unusual flight attitudes (IR)* - CFI
   - ☐☐☐ Additional items at CFI's discretion ______________________________

   ______________________________

   *IR means instrument references only
5. Postflight critique and preview of next lesson

### Completion Standards

The lesson will have been successfully completed when the pilot demonstrates at least a private pilot skill level in the assigned tasks.

**Instructor's comments:** ______________________________

______________________________

**Lesson assignment:** ______________________________

______________________________

**Notes:** ______________________________

______________________________

______________________________

______________________________

______________________________

## FLIGHT LESSON 3: SOLO PRACTICE

### Objective

To further develop the pilot's proficiency of assigned maneuvers.

### Text References

*Commercial Pilot Flight Maneuvers and Practical Test Prep* (FM)
*Pilot Handbook* (PH)
Pilot's Operating Handbook (POH)

### Content

1. Flight Lesson 2 complete? Yes ____ Copy of lesson placed in pilot's folder? Yes ____
2. Preflight briefing
3. Review items
   - ☐☐☐ Soft-field takeoff and climb - FM 197-202; POH-4
   - ☐☐☐ Soft-field approach and landing - FM 203-207; POH-4
   - ☐☐☐ Short-field takeoff and maximum performance climb - FM 208-214; POH-4
   - ☐☐☐ Short-field approach and landing - FM 215-221; POH-4
   - ☐☐☐ Power-off 180° accuracy approach and landing - FM 222-230
   - ☐☐☐ Maneuvering during slow flight - FM 284-288
   - ☐☐☐ Power-off stalls - FM 289-294; PH 71-75
   - ☐☐☐ Power-on stalls - FM 295-300; PH 71-75
   - ☐☐☐ Maneuvers as assigned by the instructor - CFI
   - ☐☐☐ Additional items at CFI's discretion ________________________________________

     ________________________________________
4. Postflight critique and preview of next lesson

### Completion Standards

The lesson will have been successfully completed when the pilot completes the solo flight. The pilot will gain proficiency in smoothness and coordination during all of the maneuvers.

**Instructor's comments:** ________________________________________

________________________________________

**Lesson assignment:** ________________________________________

________________________________________

**Notes:** ________________________________________

________________________________________

________________________________________

________________________________________

________________________________________

________________________________________

________________________________________

________________________________________

________________________________________

________________________________________

________________________________________

________________________________________

## FLIGHT LESSON 4: MANEUVERS REVIEW

### Objective

To review procedures and maneuvers covered previously.

### Text References

*Commercial Pilot Flight Maneuvers and Practical Test Prep* (FM)
*Pilot Handbook* (PH)
Pilot's Operating Handbook (POH)

### Content

1. Flight Lesson 3 complete? Yes ___ Copy of lesson placed in pilot's folder? Yes ___
2. Preflight briefing
3. Review items
   - ☐☐☐ Soft-field takeoff and climb - FM 197-202; POH-4
   - ☐☐☐ Soft-field approach and landing - FM 203-207; POH-4
   - ☐☐☐ Short-field takeoff and maximum performance climb - FM 208-214; POH-4
   - ☐☐☐ Short-field approach and landing - FM 215-221; POH-4
   - ☐☐☐ Power-off 180° accuracy approach and landing - FM 222-230
   - ☐☐☐ Traffic patterns - FM 153-159
   - ☐☐☐ Runway incursion avoidance - FM 139-143
   - ☐☐☐ Go-around/rejected landing - FM 231-235; POH-4
   - ☐☐☐ Power-off stalls - FM 289-294; PH 71-75
   - ☐☐☐ Power-on stalls - FM 295-300; PH 71-75
   - ☐☐☐ Accelerated stalls - FM 301-305; PH 71-75
   - ☐☐☐ Spin awareness - FM 306-308; POH-2
   - ☐☐☐ Emergency approach and landing (simulated) - FM 314-320; POH-3
   - ☐☐☐ Straight-and-level (IR)* - CFI
   - ☐☐☐ Constant airspeed climbs and descents (IR)* - CFI
   - ☐☐☐ Turns to a heading (IR)* - CFI
   - ☐☐☐ Recovery from unusual flight attitudes (IR)* - CFI
   - ☐☐☐ Additional items at CFI's discretion ________________________________________

   ________________________________________

   *IR means instrument references only
4. Postflight critique and preview of next lesson

### Completion Standards

The lesson will have been successfully completed when the pilot demonstrates an increase in proficiency in each maneuver. The pilot will be able to maintain the desired altitude, ±100 ft.; airspeed, ±10 kt.; and heading, ±10°.

**Instructor's comments:** ________________________________________

________________________________________

**Lesson assignment:** ________________________________________

________________________________________

**Notes:** ________________________________________

________________________________________

________________________________________

________________________________________

________________________________________

________________________________________

________________________________________

## FLIGHT LESSON 5: INTRODUCTION TO COMPLEX AIRPLANES

### Objective

To familiarize the pilot with the complex airplane, its operating characteristics, the cockpit controls, and the instruments and systems. The pilot will be introduced to preflight and postflight procedures, the use of checklists, and the safety precautions to be followed.

### Text References

*Commercial Pilot Flight Maneuvers and Practical Test Prep* (FM)
Pilot's Operating Handbook (POH)

### Content

1. Flight Lesson 4 complete? Yes ____ Copy of lesson placed in pilot's folder? Yes ____
2. Preflight briefing
3. New items
   - ☐☐☐ Certificates and documents - FM 68-71
   - ☐☐☐ Airworthiness requirements - FM 72-76
     - ☐☐☐ Minimum equipment list (MEL) - FM 73-74; CFI
   - ☐☐☐ Performance and limitations - FM 102-105; POH-2, 5, 6
   - ☐☐☐ Operation of systems - FM 106-110; POH-1, 7, 8, 9
   - ☐☐☐ Supplemental oxygen - FM 326-327; POH-7, 8
   - ☐☐☐ Pressurization - FM 328-334; POH-7
   - ☐☐☐ Use of checklists - FM 11-12, 121; POH-4
   - ☐☐☐ Preflight inspection - FM 118-122; POH-4
   - ☐☐☐ Cockpit management - FM 123-126
   - ☐☐☐ Engine starting - FM 127-131; POH-4
   - ☐☐☐ Taxiing - FM 132-138
   - ☐☐☐ Runway incursion avoidance - FM 139-143
   - ☐☐☐ Before-takeoff check - FM 144-147; POH-4
   - ☐☐☐ Normal and crosswind takeoff and climb - FM 170-178; POH-4
   - ☐☐☐ Normal and crosswind approach and landing - FM 179-196; POH-4
   - ☐☐☐ Traffic patterns - FM 153-159
   - ☐☐☐ After landing, parking, and securing - FM 336-339; POH-4
   - ☐☐☐ Additional items at CFI's discretion ______________________________

     ______________________________
4. Postflight critique and preview of next lesson

### Completion Standards

The lesson will have been successfully completed when the pilot displays an understanding of the airplane's systems, preflight procedures, and postflight procedures. The pilot will be able to demonstrate at least a private pilot skill level during the flight operations. Additionally, the pilot will display an understanding of supplemental oxygen requirements and pressurization systems and controls even if these items are not applicable to the complex airplane being flown.

**Instructor's comments:** ______________________________

______________________________

**Lesson assignment:** ______________________________

______________________________

**Notes:** ______________________________

______________________________

______________________________

______________________________

______________________________

______________________________

## FLIGHT LESSON 6: COMPLEX AIRPLANE--SLOW FLIGHT AND STALLS

### Objective

To improve the pilot's proficiency in the operation of a complex airplane and to introduce slow flight, stalls, short-field takeoffs and landings, and power-off 180° accuracy approaches and landings.

### Text References

*Commercial Pilot Flight Maneuvers and Practical Test Prep* (FM)
*Pilot Handbook* (PH)
Pilot's Operating Handbook (POH)

### Content

1. Flight Lesson 5 complete? Yes ____ Copy of lesson placed in pilot's folder? Yes ____
2. Preflight briefing
3. Review items
   - ☐☐☐ Operation of systems - FM 106-110; POH-1, 7, 8, 9
   - ☐☐☐ Supplemental oxygen - FM 326-327; POH-7, 8
   - ☐☐☐ Pressurization - FM 328-334; POH-7
   - ☐☐☐ Use of checklists - FM 11-12, 121; POH-4
   - ☐☐☐ Preflight inspection - FM 118-122; POH-4
   - ☐☐☐ Cockpit management - FM 123-126
   - ☐☐☐ Engine starting - FM 127-131; POH-4
   - ☐☐☐ Taxiing - FM 132-138
   - ☐☐☐ Runway incursion avoidance - FM 139-143
   - ☐☐☐ Before-takeoff check - FM 144-147; POH-4
   - ☐☐☐ After landing, parking, and securing - FM 336-339; POH-4

   ____________________
4. New items
   - ☐☐☐ Short-field takeoff and maximum performance climb - FM 208-214; POH-4
   - ☐☐☐ Short-field approach and landing - FM 215-221; POH-4
   - ☐☐☐ Power-off 180° accuracy approach and landing - FM 222-230
   - ☐☐☐ Maneuvering during slow flight - FM 284-288
   - ☐☐☐ Power-off stalls - FM 289-294; PH 71-75
   - ☐☐☐ Power-on stalls - FM 295-300; PH 71-75
   - ☐☐☐ Accelerated stalls - FM 301-305; PH 71-75
   - ☐☐☐ Spin awareness - FM 306-308; POH-2
   - ☐☐☐ Additional items at CFI's discretion ____________________

   ____________________
5. Postflight critique and preview of next lesson

### Completion Standards

The lesson will have been successfully completed when the pilot displays an increased proficiency in the operation of a complex airplane. During this and subsequent flights, the pilot will perform the preflight inspection, engine starting, taxiing, the before-takeoff check, and the postflight procedures without instructor assistance. The pilot will be able to demonstrate an understanding of the short-field takeoffs and landings, power-off accuracy approaches and landings, slow flight, and stalls (including the proper recovery procedures).

**Instructor's comments:** ____________________

____________________

**Lesson assignment:** ____________________

____________________

**Notes:** ____________________

____________________

____________________

## FLIGHT LESSON 7: COMPLEX AIRPLANE--EMERGENCY OPERATIONS

### Objective

To improve the pilot's proficiency in the operation of a complex airplane and to introduce emergency operations and soft-field takeoffs and landings.

### Text References

*Commercial Pilot Flight Maneuvers and Practical Test Prep* (FM)
*Pilot Handbook* (PH)
Pilot's Operating Handbook (POH)

### Content

1. Flight Lesson 6 complete? Yes ____ Copy of lesson placed in pilot's folder? Yes ____
2. Preflight briefing
3. Review items
   - ☐☐☐ Operation of systems - FM 106-110; POH-1, 7, 8, 9
   - ☐☐☐ Short-field takeoff and maximum performance climb - FM 208-214; POH-4
   - ☐☐☐ Short-field approach and landing - FM 215-221; POH-4
   - ☐☐☐ Power-off 180° accuracy approach and landing - FM 222-230
   - ☐☐☐ Maneuvering during slow flight - FM 284-288
   - ☐☐☐ Power-off stalls - FM 289-294; PH 71-75
   - ☐☐☐ Power-on stalls - FM 295-300; PH 71-75
   - ☐☐☐ Accelerated stalls - FM 301-305; PH 71-75
   - ☐☐☐ Spin awareness - FM 306-308; POH-2

   ________________________________________
4. New items
   - ☐☐☐ Soft-field takeoff and climb - FM 197-202; POH-4
   - ☐☐☐ Soft-field approach and landing - FM 203-207; POH-4
   - ☐☐☐ Go-around/rejected landing - FM 231-235; POH-4
   - ☐☐☐ Emergency descent - FM 311-313; POH-3
   - ☐☐☐ Emergency approach and landing (simulated) - FM 314-320; POH-3
   - ☐☐☐ Systems and equipment malfunctions - FM 321-322; POH-3, 9
   - ☐☐☐ Emergency equipment and survival gear - FM 323-324; POH-7, 9
   - ☐☐☐ Additional items at CFI's discretion ________________________________________

   ________________________________________
5. Postflight critique and preview of next lesson

### Completion Standards

The lesson will have been successfully completed when the pilot displays proficiency in short-field takeoffs and landings, power-off accuracy approaches and landings, slow flight, and recovery from stalls. Additionally, the pilot will display an understanding of emergency procedures, go-arounds, and soft-field takeoffs and landings.

**Instructor's comments:** ________________________________________

________________________________________

**Lesson assignment:** ________________________________________

________________________________________

**Notes:** ________________________________________

________________________________________

________________________________________

________________________________________

________________________________________

________________________________________

## FLIGHT LESSON 8: COMPLEX AIRPLANE--NIGHT OPERATIONS

### Objective

To introduce the pilot to night-flying preparation and night-flying operations to the commercial pilot skill level. Additionally, the pilot will demonstrate the necessary skills and proficiency to act as pilot in command of a complex airplane.

### Text References

*Commercial Pilot Flight Maneuvers and Practical Test Prep* (FM)
*Pilot Handbook* (PH)
Pilot's Operating Handbook (POH)

### Content

1. Flight Lesson 7 complete? Yes ___ Copy of lesson placed in pilot's folder? Yes ___
2. Preflight briefing
3. New items (night operations)
   - ☐☐☐ Operation of systems - FM 106-110; POH-1, 7, 8, 9
   - ☐☐☐ Performance and limitations - FM 102-105; POH-2, 5, 6
   - ☐☐☐ Aeromedical factors - FM 111-116
   - ☐☐☐ Physiological aspects of night flying - CFI
   - ☐☐☐ Lighting and equipment for night flight - CFI
   - ☐☐☐ Preflight inspection - FM 118-122; POH-4
   - ☐☐☐ Radio communications and ATC light signals - FM 150-152
   - ☐☐☐ Traffic patterns - FM 153-159
   - ☐☐☐ Airport, runway, and taxiway signs, markings, and lighting - FM 160-167
   - ☐☐☐ Runway incursion avoidance - FM 139-143
   - ☐☐☐ Soft-field takeoff and climb - FM 197-202; POH-4
   - ☐☐☐ Soft-field approach and landing - FM 203-207; POH-4
   - ☐☐☐ Short-field takeoff and maximum performance climb - FM 208-214; POH-4
   - ☐☐☐ Short-field approach and landing - FM 215-221; POH-4
   - ☐☐☐ Go-around/rejected landing - FM 231-235; POH-4
   - ☐☐☐ Maneuvering during slow flight - FM 284-288
   - ☐☐☐ Power-off stalls - FM 289-294; PH 71-75
   - ☐☐☐ Power-on stalls - FM 295-300; PH 71-75
   - ☐☐☐ Accelerated stalls - FM 301-305; PH 71-75
   - ☐☐☐ Emergency descent - FM 311-313; POH-3
   - ☐☐☐ Emergency approach and landing (simulated) - FM 314-320; POH-3
   - ☐☐☐ Systems and equipment malfunctions - FM 321-322; POH-3, 9
   - ☐☐☐ Additional items at CFI's discretion ________________________________
     ________________________________________________
4. Postflight critique and preview of next lesson

### Completion Standards

The lesson will have been successfully completed when the pilot demonstrates proficiency in the operation of a complex airplane as pilot in command. The pilot will be able to maintain the desired altitude, ±100 ft.; airspeed, ±10 kt.; and heading, ±10°.

**Instructor's comments:** ________________________________________________

________________________________________________

**Lesson assignment:** ________________________________________________

________________________________________________

**Notes:** ________________________________________________

________________________________________________

________________________________________________

________________________________________________

## FLIGHT LESSON 9: SOLO PRACTICE

### Objective

To further develop the pilot's confidence and proficiency through solo practice of assigned maneuvers.

### Text References

*Commercial Pilot Flight Maneuvers and Practical Test Prep* (FM)
*Pilot Handbook* (PH)
Pilot's Operating Handbook (POH)

### Content

1. Flight Lesson 8 complete? Yes ____ Copy of lesson placed in pilot's folder? Yes ____
2. Preflight briefing
3. Review items
   - ☐☐☐ Soft-field takeoff and climb - FM 197-202; POH-4
   - ☐☐☐ Soft-field approach and landing - FM 203-207; POH-4
   - ☐☐☐ Short-field takeoff and maximum performance climb - FM 208-214; POH-4
   - ☐☐☐ Short-field approach and landing - FM 215-221; POH-4
   - ☐☐☐ Power-off 180° accuracy approach and landing - FM 222-230
   - ☐☐☐ Traffic patterns - FM 153-159
   - ☐☐☐ Maneuvering during slow flight - FM 284-288
   - ☐☐☐ Power-off stalls - FM 289-294; PH 71-75
   - ☐☐☐ Power-on stalls - FM 295-300; PH 71-75
   - ☐☐☐ Accelerated stalls - FM 301-305; PH 71-75
   - ☐☐☐ Maneuvers as assigned by the instructor - CFI
   - ☐☐☐ Additional items at CFI's discretion ____________________

     ____________________
4. Postflight critique and preview of next lesson

### Completion Standards

The lesson will have been successfully completed when the pilot completes the listed maneuvers assigned for this solo flight. The pilot will gain confidence and proficiency as a result of the solo practice.

**Instructor's comments:** ____________________

____________________

**Lesson assignment:** ____________________

____________________

**Notes:** ____________________

____________________

____________________

____________________

____________________

____________________

____________________

____________________

____________________

____________________

## FLIGHT LESSON 10: INTRODUCTION TO PERFORMANCE MANEUVERS

### Objective

To introduce the pilot to steep turns, steep spirals, chandelles, and lazy eights.

### Text References

*Commercial Pilot Flight Maneuvers and Practical Test Prep* (FM)
Pilot's Operating Handbook (POH)

### Content

1. Flight Lesson 9 complete? Yes ___ Copy of lesson placed in pilot's folder? Yes ___
2. Preflight briefing
3. Review items
   - ☐☐☐ Performance and limitations - FM 102-105; POH-2, 5, 6
   - ☐☐☐ Soft-field takeoff and climb - FM 197-202; POH-4
   - ☐☐☐ Soft-field approach and landing - FM 203-207; POH-4
   - ☐☐☐ Go-around/rejected landing - FM 231-235; POH-4
   - ☐☐☐ Collision avoidance procedures - FM 153-154

   ______________________________
4. New items
   - ☐☐☐ Steep turns - FM 238-242
   - ☐☐☐ Steep spirals - FM 243-248
   - ☐☐☐ Chandelles - FM 249-253
   - ☐☐☐ Lazy eights - FM 254-259
   - ☐☐☐ Additional items at CFI's discretion ______________________________

   ______________________________
5. Postflight critique and preview of next lesson

### Completion Standards

The lesson will have been successfully completed when the pilot demonstrates the proper entry procedures and understands the control techniques required for steep turns, steep spirals, chandelles, and lazy eights. The pilot will be able to maintain the desired altitude, ±100 ft.; airspeed, ±10 kt.; and heading, ±10°.

**Instructor's comments:** ______________________________

______________________________

**Lesson assignment:** ______________________________

______________________________

**Notes:** ______________________________

______________________________

______________________________

______________________________

______________________________

______________________________

______________________________

______________________________

______________________________

______________________________

______________________________

## FLIGHT LESSON 11: INTRODUCTION TO EIGHTS-ON-PYLONS

### Objective

To introduce the pilot to eights-on-pylons.

### Text References

*Commercial Pilot Flight Maneuvers and Practical Test Prep* (FM)
Pilot's Operating Handbook (POH)

### Content

1. Flight Lesson 10 complete? Yes ____ Copy of lesson placed in pilot's folder? Yes ____
2. Preflight briefing
3. Review items
   - ☐☐☐ Short-field takeoff and maximum performance climb - FM 208-214; POH-4
   - ☐☐☐ Short-field approach and landing - FM 215-221; POH-4
   - ☐☐☐ Power-off 180° accuracy approach and landing - FM 222-230
   - ☐☐☐ Steep turns - FM 238-242
   - ☐☐☐ Steep spirals - FM 243-248
   - ☐☐☐ Chandelles - FM 249-253
   - ☐☐☐ Lazy eights - FM 254-259
   - ☐☐☐ Emergency descent - FM 311-313; POH-3
   - ☐☐☐ Emergency approach and landing (simulated) - FM 314-320; POH-3
   - ☐☐☐ Traffic patterns - FM 153-159
     - ☐☐☐ Runway incursion and collision avoidance - FM 153-154
     - ☐☐☐ Wake turbulence avoidance - FM 154-155
     - ☐☐☐ Wind shear avoidance - FM 155

   ____________________________________________
4. New item
   - ☐☐☐ Eights-on-pylons - FM 262-267
   - ☐☐☐ Additional items at CFI's discretion ____________________________________________

   ____________________________________________
5. Postflight critique and preview of next lesson

### Completion Standards

The lesson will have been successfully completed when the pilot demonstrates the proper entry procedures and understands how to keep the line-of-sight reference line correctly on the pylon during the performance of eights-on-pylons. The pilot will be able to maintain the desired altitude, ±100 ft.; airspeed, ±10 kt.; and heading, ±10°.

**Instructor's comments:** ____________________________________________

____________________________________________

**Lesson assignment:** ____________________________________________

____________________________________________

**Notes:** ____________________________________________

____________________________________________

____________________________________________

____________________________________________

____________________________________________

____________________________________________

____________________________________________

____________________________________________

## FLIGHT LESSON 12: SOLO PRACTICE

### Objective

To develop the pilot's confidence and proficiency through solo practice of the assigned maneuvers.

### Text References

*Commercial Pilot Flight Maneuvers and Practical Test Prep* (FM)
Pilot's Operating Handbook (POH)

### Content

1. Flight Lesson 11 complete? Yes ____ Copy of lesson placed in pilot's folder? Yes ____
2. Preflight briefing
3. Review items
   - ☐☐☐ Preflight procedures - FM 118-147; POH-4
   - ☐☐☐ Soft-field takeoff and climb - FM 197-202; POH-4
   - ☐☐☐ Soft-field approach and landing - FM 203-207; POH-4
   - ☐☐☐ Steep turns - FM 238-242
   - ☐☐☐ Steep spirals - FM 243-248
   - ☐☐☐ Chandelles - FM 249-253
   - ☐☐☐ Lazy eights - FM 254-259
   - ☐☐☐ Eights-on-pylons - FM 262-267
   - ☐☐☐ After landing, parking, and securing - FM 336-339; POH-4
   - ☐☐☐ Additional items at CFI's discretion ____________________
4. Postflight critique and preview of next lesson

### Completion Standards

The lesson will have been successfully completed when the pilot completes the listed maneuvers assigned for the solo flight. The pilot will gain confidence and proficiency as a result of the solo practice.

**Instructor's comments:** ____________________

**Lesson assignment:** ____________________

**Notes:** ____________________

## FLIGHT LESSON 13: SOLO PRACTICE

### Objective

To further develop the pilot's proficiency through solo practice of the assigned maneuvers.

### Text References

*Commercial Pilot Flight Maneuvers and Practical Test Prep* (FM)
*Pilot Handbook* (PH)
Pilot's Operating Handbook (POH)

### Content

1. Flight Lesson 12 complete? Yes ____ Copy of lesson placed in pilot's folder? Yes ____
2. Preflight briefing
3. Review items
   - ☐☐☐ Short-field takeoff and maximum performance climb - FM 208-214; POH-4
   - ☐☐☐ Short-field approach and landing - FM 215-221; POH-4
   - ☐☐☐ Power-off 180° accuracy approach and landing - FM 222-230
   - ☐☐☐ Steep turns - FM 238-242
   - ☐☐☐ Steep spirals - FM 243-248
   - ☐☐☐ Chandelles - FM 249-253
   - ☐☐☐ Lazy eights - FM 254-259
   - ☐☐☐ Eights-on-pylons - FM 262-267
   - ☐☐☐ Maneuvering during slow flight - FM 284-288
   - ☐☐☐ Power-off stalls - FM 289-294; PH 71-75
   - ☐☐☐ Power-on stalls - FM 295-300; PH 71-75
   - ☐☐☐ Accelerated stalls - FM 301-305; PH 71-75
   - ☐☐☐ Emergency descent - FM 311-313; POH-3
   - ☐☐☐ Additional items at CFI's discretion ____________________
4. Postflight critique and preview of next lesson

### Completion Standards

The lesson will have been successfully completed when the pilot completes the listed maneuvers assigned for the solo flight. The pilot will gain proficiency as a result of the solo practice.

**Instructor's comments:** ____________________

**Lesson assignment:** ____________________

**Notes:** ____________________

## FLIGHT LESSON 14: MANEUVERS REVIEW

### Objective

To review procedures and maneuvers covered previously.

### Text References

*Commercial Pilot Flight Maneuvers and Practical Test Prep* (FM)
*Pilot Handbook* (PH)
Pilot's Operating Handbook (POH)

### Content

1. Flight Lesson 13 complete? Yes ____ Copy of lesson placed in pilot's folder? Yes ____
2. Preflight briefing
3. Review items
   - ☐☐☐ Radio communications and ATC light signals - FM 150-152
   - ☐☐☐ Traffic patterns - FM 153-159
   - ☐☐☐ Airport, runway, and taxiway signs, markings, and lighting - FM 160-167
   - ☐☐☐ Runway incursion avoidance - FM 139-143
   - ☐☐☐ Soft-field takeoff and climb - FM 197-202; POH-4
   - ☐☐☐ Soft-field approach and landing - FM 203-207; POH-4
   - ☐☐☐ Short-field takeoff and maximum performance climb - FM 208-214; POH-4
   - ☐☐☐ Short-field approach and landing - FM 215-221; POH-4
   - ☐☐☐ Power-off 180° accuracy approach and landing - FM 222-230
   - ☐☐☐ Go-around/rejected landing - FM 231-235; POH-4
   - ☐☐☐ Steep turns - FM 238-242
   - ☐☐☐ Steep spirals - FM 243-248
   - ☐☐☐ Chandelles - FM 249-253
   - ☐☐☐ Lazy eights - FM 254-259
   - ☐☐☐ Eights-on-pylons - FM 262-267
   - ☐☐☐ Maneuvering during slow flight - FM 284-288
   - ☐☐☐ Power-off stalls - FM 289-294; PH 71-75
   - ☐☐☐ Power-on stalls - FM 295-300; PH 71-75
   - ☐☐☐ Accelerated stalls - FM 301-305; PH 71-75
   - ☐☐☐ Straight-and-level (IR) - CFI
   - ☐☐☐ Constant airspeed climbs and descents (IR) - CFI
   - ☐☐☐ Turns to a heading (IR) - CFI
   - ☐☐☐ Recovery from unusual attitudes (IR) - CFI
   - ☐☐☐ Additional items at CFI's discretion ____________________

     ____________________
4. Postflight critique and preview of next lesson

### Completion Standards

The lesson will have been successfully completed when the pilot demonstrates an increase in proficiency in each maneuver. The pilot will be able to maintain the desired altitude, ±100 ft.; airspeed, ±10 kt.; and heading, ±10°. During the takeoffs, the pilot will be able to maintain $V_X$, +10/–0 kt., and $V_Y$, ±5 kt., as appropriate. During the approach for a landing, the pilot will be able to maintain the recommended airspeed, ±5 kt.

**Instructor's comments:** ____________________

____________________

**Lesson assignment:** ____________________

____________________

**Notes:** ____________________

____________________

____________________

____________________

## FLIGHT LESSON 15: COMPLEX AIRPLANE--MANEUVERS REVIEW

### Objective

To review procedures and maneuvers covered previously in a complex airplane.

### Text References

*Commercial Pilot Flight Maneuvers and Practical Test Prep* (FM)
*Pilot Handbook* (PH)
Pilot's Operating Handbook (POH)

### Content

1. Flight Lesson 14 complete? Yes ____ Copy of lesson placed in pilot's folder? Yes ____
2. Preflight briefing
3. Review items
   - ☐☐☐ Certificates and documents - FM 68-71
   - ☐☐☐ Airworthiness requirements - FM 72-76
     - ☐☐☐ Minimum equipment list (MEL) - FM 73-74; CFI
   - ☐☐☐ Performance and limitations - FM 102-105; POH-2, 5, 6
     - ☐☐☐ Operations at maximum takeoff weight - FM 103
     - ☐☐☐ High density altitude operations - FM 104-105
   - ☐☐☐ Operation of systems - FM 106-110; POH-1, 7, 8, 9
   - ☐☐☐ Supplemental oxygen - FM 326-327; POH-7, 8
   - ☐☐☐ Pressurization - FM 328-334; POH-7
   - ☐☐☐ Aeromedical factors - FM 111-116
   - ☐☐☐ Aeronautical decision making (ADM) - PH 343-349
   - ☐☐☐ Traffic patterns - FM 153-159
   - ☐☐☐ Soft-field takeoff and climb - FM 197-202; POH-4
   - ☐☐☐ Soft-field approach and landing - FM 203-207; POH-4
   - ☐☐☐ Short-field takeoff and maximum performance climb - FM 208-214; POH-4
   - ☐☐☐ Short-field approach and landing - FM 215-221; POH-4
   - ☐☐☐ Power-off 180° accuracy approach and landing - FM 222-230
   - ☐☐☐ Go-around/rejected landing - FM 231-235; POH-4
   - ☐☐☐ Emergency descent - FM 311-313; POH-3
   - ☐☐☐ Emergency approach and landing (simulated) - FM 314-320; POH-3
   - ☐☐☐ Systems and equipment malfunctions - FM 321-322; POH-3, 9
   - ☐☐☐ Maneuvering during slow flight - FM 284-288
   - ☐☐☐ Power-off stalls - FM 289-294; PH 71-75
   - ☐☐☐ Power-on stalls - FM 295-300; PH 71-75
   - ☐☐☐ Accelerated stalls - FM 301-305; PH 71-75
   - ☐☐☐ Spin awareness - FM 306-308; POH-2
   - ☐☐☐ Additional items at CFI's discretion ________________________________

     ________________________________
4. Postflight critique and preview of next lesson

### Completion Standards

The lesson will have been successfully completed when the pilot demonstrates an increase in proficiency in each maneuver while operating a complex airplane. While maneuvering during slow flight, the pilot will be able to maintain the desired airspeed, +5/–0 kt.; specified altitude, ±100 ft.; heading, ±10°; specified bank angle, ±10°, during turning flight; and roll out on the specified heading, ±10°.

**Instructor's comments:** ________________________________

________________________________

**Lesson assignment:** ________________________________

________________________________

**Notes:** ________________________________

________________________________

## FLIGHT LESSON 16: SOLO PRACTICE

### Objective

To further develop the pilot's proficiency through solo practice of the assigned maneuvers. This flight provides the pilot the opportunity to prepare for the Stage One Check.

### Text References

*Commercial Pilot Flight Maneuvers and Practical Test Prep* (FM)
*Pilot Handbook* (PH)
Pilot's Operating Handbook (POH)

### Content

1. Flight Lesson 15 complete? Yes ____ Copy of lesson placed in pilot's folder? Yes ____
2. Preflight briefing
3. Review items
   - ☐☐☐ Normal and crosswind takeoff and climb - FM 170-178; POH-4
   - ☐☐☐ Normal and crosswind approach and landing - FM 179-196; POH-4
   - ☐☐☐ Soft-field takeoff and climb - FM 197-202; POH-4
   - ☐☐☐ Soft-field approach and landing - FM 203-207; POH-4
   - ☐☐☐ Short-field takeoff and maximum performance climb - FM 208-214; POH-4
   - ☐☐☐ Short-field approach and landing - FM 215-221; POH-4
   - ☐☐☐ Power-off 180° accuracy approach and landing - FM 222-230
   - ☐☐☐ Maneuvering during slow flight - FM 284-288
   - ☐☐☐ Power-off stalls - FM 289-294; PH 71-75
   - ☐☐☐ Power-on stalls - FM 295-300; PH 71-75
   - ☐☐☐ Accelerated stalls - FM 301-305; PH 71-75
   - ☐☐☐ Maneuvers as assigned by the instructor - CFI
   - ☐☐☐ Additional items at CFI's discretion ____________________

     ____________________
4. Postflight critique and preview of next lesson

### Completion Standards

The lesson will have been successfully completed when the pilot completes the listed maneuvers assigned for this solo flight. The pilot's proficiency will increase as a result of the solo practice.

**Instructor's comments:** ____________________

____________________

**Lesson assignment:** ____________________

____________________

**Notes:** ____________________

____________________

____________________

____________________

____________________

____________________

____________________

____________________

____________________

____________________

____________________

## FLIGHT LESSON 17: STAGE ONE CHECK--COMPLEX AIRPLANE

### Objective

During this stage check, an authorized flight instructor will evaluate the pilot's proficiency in a complex airplane.

### Text References

*Commercial Pilot Flight Maneuvers and Practical Test Prep* (FM)
*Pilot Handbook* (PH)
Pilot's Operating Handbook (POH)

### Content

1. Flight Lesson 16 complete? Yes ____ Copy of lesson placed in pilot's folder? Yes ____
2. Preflight briefing
3. Stage check tasks
   - ☐☐☐ Certificates and documents - FM 68-71
   - ☐☐☐ Airworthiness requirements - FM 72-76
     - ☐☐☐ Minimum equipment list (MEL) - FM 73-74; CFI
   - ☐☐☐ Performance and limitations - FM 102-105; POH-2, 5, 6
   - ☐☐☐ Operation of systems - FM 106-110; POH-1, 7, 8, 9
   - ☐☐☐ Supplemental oxygen - FM 326-327; POH-7, 8
   - ☐☐☐ Pressurization - FM 328-334; POH-7
   - ☐☐☐ Aeromedical factors - FM 111-116
   - ☐☐☐ Aeronautical decision making (ADM) - PH 343-349
   - ☐☐☐ Preflight inspection - FM 118-122; POH-4
   - ☐☐☐ Cockpit management - FM 123-126
   - ☐☐☐ Runway incursion avoidance - FM 139-143
   - ☐☐☐ Soft-field takeoff and climb - FM 197-202; POH-4
   - ☐☐☐ Soft-field approach and landing - FM 203-207; POH-4
   - ☐☐☐ Short-field takeoff and maximum performance climb - FM 208-214; POH-4
   - ☐☐☐ Short-field approach and landing - FM 215-221; POH-4
   - ☐☐☐ Power-off 180° accuracy approach and landing - FM 222-230
   - ☐☐☐ Go-around/rejected landing - FM 231-235; POH-4
   - ☐☐☐ Maneuvering during slow flight - FM 284-288
   - ☐☐☐ Power-off stalls - FM 289-294; PH 71-75
   - ☐☐☐ Power-on stalls - FM 295-300; PH 71-75
   - ☐☐☐ Accelerated stalls - FM 301-305; PH 71-75
   - ☐☐☐ Straight-and-level (IR) - CFI
   - ☐☐☐ Constant airspeed climbs and descents (IR) - CFI
   - ☐☐☐ Turns to a heading (IR) - CFI
   - ☐☐☐ Recovery from unusual attitudes (IR) - CFI
   - ☐☐☐ Emergency descent - FM 311-313; POH-3
   - ☐☐☐ Emergency approach and landing (simulated) - FM 314-320; POH-3
   - ☐☐☐ Systems and equipment malfunctions - FM 321-322; POH-3, 9
   - ☐☐☐ After landing, parking, and securing - FM 336-339; POH-4
   - ☐☐☐ Additional items at CFI's discretion ____________________
4. Postflight critique and preview of next lesson

### Completion Standards

The lesson and Stage One will have been successfully completed when the pilot can demonstrate proficiency in the operation and systems of the complex airplane and receive a complex airplane logbook endorsement (if applicable). While maneuvering during slow flight, the pilot will be able to maintain the desired airspeed, +5/–0 kt. While performing all maneuvers, the pilot will be able to maintain the specified altitude, ±100 ft.; heading, ±10°; specified bank angle, ±10°, during turning flight; and roll out on the specified heading, ±10°.

**Instructor's comments:** ____________________

**Lesson assignment:** ____________________

**Notes:** ____________________

____________________

## STAGE TWO

### Stage Two Objective

The pilot will increase his/her skill and proficiency in both day and night cross-country operations.

### Stage Two Completion Standards

The stage will be completed when the pilot demonstrates the ability to plan and conduct cross-country flights in an airplane while operating under day and night conditions using pilotage, dead reckoning, and navigation systems.

| Lesson | Topic |
|---|---|
| 18 | Maneuvers Review |
| 19 | Dual Cross-Country |
| 20 | Night Flight--Local |
| 21 | Dual Night Cross-Country |
| 22 | Solo Night--Local |
| 23 | Solo Night Cross-Country |
| 24 | Solo Cross-Country |
| 25 | Instrument Review |
| 26 | Solo Cross-Country |
| 27 | Solo Cross-Country |
| 28 | Solo Night--Local |
| 29 | Solo Cross-Country |
| 30 | Solo Cross-Country |
| 31 | Maneuvers Review |
| 32 | Solo Cross-Country |
| 33 | Solo Cross-Country |
| 34 | Stage Two Check |

## FLIGHT LESSON 18: MANEUVERS REVIEW

### Objective

To review procedures and maneuvers covered previously.

### Text References

*Commercial Pilot Flight Maneuvers and Practical Test Prep* (FM)
*Pilot Handbook* (PH)
Pilot's Operating Handbook (POH)

### Content

1. Flight Lesson 17 complete? Yes ____ Copy of lesson placed in pilot's folder? Yes ____
2. Preflight briefing
3. Review items
   - ☐☐☐ Short-field takeoff and maximum performance climb - FM 208-214; POH-4
   - ☐☐☐ Short-field approach and landing - FM 215-221; POH-4
   - ☐☐☐ Power-off 180° accuracy approach and landing - FM 222-230
   - ☐☐☐ Collision avoidance procedures - FM 153-154
   - ☐☐☐ Wake turbulence avoidance - FM 154-155
   - ☐☐☐ Go-around/rejected landing - FM 231-235; POH-4
   - ☐☐☐ Maneuvering during slow flight - FM 284-288
   - ☐☐☐ Power-off stalls - FM 289-294; PH 71-75
   - ☐☐☐ Power-on stalls - FM 295-300; PH 71-75
   - ☐☐☐ Accelerated stalls - FM 301-305; PH 71-75
   - ☐☐☐ Spin awareness - FM 306-308; POH-2
   - ☐☐☐ Emergency descent - FM 311-313; POH-3
   - ☐☐☐ Emergency approach and landing (simulated) - FM 314-320; POH-3
   - ☐☐☐ Straight-and-level (IR) - CFI
   - ☐☐☐ Maneuvering during slow flight (IR) - CFI
   - ☐☐☐ Power-off stall (IR) - CFI
   - ☐☐☐ Power-on stall (IR) - CFI
   - ☐☐☐ Constant airspeed climbs and descents (IR) - CFI
   - ☐☐☐ Turns to a heading (IR) - CFI
   - ☐☐☐ Additional items at CFI's discretion ________________________________________

     ________________________________________
4. Postflight critique and preview of next lesson

### Completion Standards

The lesson will have been successfully completed when the pilot demonstrates an increase in proficiency in each maneuver. The pilot will be able to maintain the desired altitude, ±100 ft.; airspeed, ±10 kt.; and heading, ±10°. Additionally, during the short-field takeoff and climb, the pilot will be able to maintain $V_X$, +10/–0 kt., and $V_Y$, ±10 kt.

**Instructor's comments:** ________________________________________

________________________________________

**Lesson assignment:** ________________________________________

________________________________________

**Notes:** ________________________________________

________________________________________

________________________________________

________________________________________

________________________________________

________________________________________

## FLIGHT LESSON 19: DUAL CROSS-COUNTRY

### Objective

To introduce the pilot to planning and executing a daytime cross-country flight to the commercial pilot skill level. This flight must be at least 2 hours in duration, a total straight-line distance of more than 100 NM from the original point of departure, and occur in daytime conditions.

### Text References

*Commercial Pilot Flight Maneuvers and Practical Test Prep* (FM)
*Pilot Handbook* (PH)
Pilot's Operating Handbook (POH)

### Content

1. Flight Lesson 18 complete? Yes ____ Copy of lesson placed in pilot's folder? Yes ____
2. Preflight briefing
3. Review items

☐☐☐ Certificates and documents - FM 68-71
☐☐☐ Performance and limitations - FM 102-105; POH-2, 5, 6
☐☐☐ Aeromedical factors - FM 111-116
☐☐☐ Aeronautical decision making (ADM) - PH 343-349
☐☐☐ Cockpit management - FM 123-126
☐☐☐ Radio communications and ATC light signals - FM 150-152
☐☐☐ Airport, runway, and taxiway signs, markings, and lighting - FM 160-167
☐☐☐ Runway incursion avoidance - FM 139-143
☐☐☐ Soft-field takeoff and climb - FM 197-202; POH-4
☐☐☐ Soft-field approach and landing - FM 203-207; POH-4
☐☐☐ Short-field takeoff and maximum performance climb - FM 208-214; POH-4
☐☐☐ Short-field approach and landing - FM 215-221; POH-4
☐☐☐ Systems and equipment malfunctions - FM 321-322; POH-3, 9
☐☐☐ Straight-and-level (IR) - CFI
☐☐☐ Constant airspeed climbs and descents (IR) - CFI
☐☐☐ Turns to a heading (IR) - CFI
☐☐☐ Recovery from unusual attitudes (IR) - CFI
☐☐☐ Partial panel flying (IR) - CFI
☐☐☐ Intercepting and tracking navigational systems (IR) - CFI

4. New items

☐☐☐ Weather information - FM 77-91
☐☐☐ Cross-country flight planning - FM 92-94
☐☐☐ National Airspace System - FM 95-101
☐☐☐ Open a flight plan - CFI
☐☐☐ Pilotage and dead reckoning - FM 270-274
☐☐☐ Magnetic compass turns - FM 272-274; PH 96-101
☐☐☐ Navigation systems - FM 275-276; PH 481-514
☐☐☐ ATC radar services - PH 199-201
☐☐☐ Diversion - FM 277-279
☐☐☐ Lost procedures - FM 280-282
☐☐☐ Closing a flight plan - CFI
☐☐☐ Additional items at CFI's discretion ____________________

5. Postflight critique and preview of next lesson

NOTE: This lesson may be completed under VFR or IFR at the discretion of the flight instructor and/or chief flight instructor, as appropriate.

### Completion Standards

The lesson will have been successfully completed when the pilot is able to plan and conduct a daytime cross-country flight. The pilot will be able to maintain altitude, ±100 ft.; heading, ±10°; and the desired airspeed, ±10 kt. Additionally, the pilot will be able to verify the airplane's position within 2 NM of the flight planned route at all times.

**Instructor's comments:** ____________________

____________________

**Lesson assignment:** ____________________

**Notes:** ____________________

____________________

## FLIGHT LESSON 20: NIGHT FLIGHT--LOCAL

### Objective

To increase the pilot's experience and proficiency in night operations.

### Text References

*Commercial Pilot Flight Maneuvers and Practical Test Prep* (FM)
*Pilot Handbook* (PH)
Pilot's Operating Handbook (POH)

### Content

1. Flight Lesson 19 complete? Yes ____ Copy of lesson placed in pilot's folder? Yes ____
2. Preflight briefing
3. Review items
   - ☐☐☐ Physiological aspects of night flying - CFI
   - ☐☐☐ Lighting and equipment for night flight - CFI
   - ☐☐☐ Airport, runway, and taxiway signs, markings, and lighting - FM 160-167
   - ☐☐☐ Preflight inspection - FM 118-122; POH-4
   - ☐☐☐ Cockpit management - FM 123-126
   - ☐☐☐ Engine starting - FM 127-131; POH-4
   - ☐☐☐ Taxiing - FM 132-138
   - ☐☐☐ Runway incursion avoidance - FM 139-143
   - ☐☐☐ Before-takeoff check - FM 144-147; POH-4
   - ☐☐☐ Radio communications and ATC light signals - FM 150-152
   - ☐☐☐ Normal and crosswind takeoff and climb - FM 170-178; POH-4
   - ☐☐☐ Short-field takeoff and maximum performance climb - FM 208-214; POH-4
   - ☐☐☐ Short-field approach and landing - FM 215-221; POH-4
   - ☐☐☐ Traffic patterns - FM 153-159
   - ☐☐☐ Go-around/rejected landing - FM 231-235; POH-4
   - ☐☐☐ Collision and runway incursion avoidance procedures - FM 153-154
   - ☐☐☐ Maneuvering during slow flight - FM 284-288
   - ☐☐☐ Power-off stalls - FM 289-294; PH 71-75
   - ☐☐☐ Power-on stalls - FM 295-300; PH 71-75
   - ☐☐☐ Accelerated stalls - FM 301-305; PH 71-75
   - ☐☐☐ Emergency approach and landing (simulated) - FM 314-320; POH-3
   - ☐☐☐ Systems and equipment malfunctions - FM 321-322; POH-3, 9
   - ☐☐☐ Emergency descent - FM 311-313; POH-3
   - ☐☐☐ Normal and crosswind approach and landing - FM 179-196; POH-4
   - ☐☐☐ Power-off 180° accuracy approach and landing - FM 222-230
   - ☐☐☐ Straight-and-level (IR) - CFI
   - ☐☐☐ Constant airspeed climbs and descents (IR) - CFI
   - ☐☐☐ Turns to a heading (IR) - CFI
   - ☐☐☐ Recovery from unusual attitudes (IR) - CFI
   - ☐☐☐ After landing, parking, and securing - FM 336-339; POH-4
   - ☐☐☐ Additional items at CFI's discretion ____________________

     ____________________
4. Postflight critique and preview of next lesson

### Completion Standards

The lesson will have been successfully completed when the pilot demonstrates the knowledge of night flying preparation, precautions, and procedures. The pilot will be able to maintain the desired altitude, ±100 ft.; airspeed, ±10 kt.; and heading, ±10°.

**Instructor's comments:** ____________________

____________________

**Lesson assignment:** ____________________

____________________

**Notes:** ____________________

____________________

## FLIGHT LESSON 21: DUAL NIGHT CROSS-COUNTRY

### Objective

To introduce the pilot to planning and executing a nighttime cross-country flight to the commercial pilot skill level. This flight must be at least 2 hours in duration, a total straight-line distance of more than 100 NM from the original point of departure, and occur in nighttime conditions.

### Text References

*Commercial Pilot Flight Maneuvers and Practical Test Prep* (FM)
*Pilot Handbook* (PH)
Pilot's Operating Handbook (POH)

### Content

1. Flight Lesson 20 complete? Yes ____ Copy of lesson placed in pilot's folder? Yes ____
2. Preflight briefing
3. Review items

☐☐☐ Physiological aspects of night flying - CFI
☐☐☐ Lighting and equipment for night flight - CFI
☐☐☐ Performance and limitations - FM 102-105; POH-2, 5, 6
☐☐☐ Airport, runway, and taxiway signs, markings, and lighting - FM 160-167
☐☐☐ Runway incursion avoidance - FM 139-143
☐☐☐ Preflight inspection - FM 118-122; POH-4
☐☐☐ Normal and crosswind takeoff and climb - FM 170-178; POH-4
☐☐☐ Normal and crosswind approach and landing - FM 179-196; POH-4
☐☐☐ Soft-field takeoff and climb - FM 197-202; POH-4
☐☐☐ Soft-field approach and landing - FM 203-207; POH-4
☐☐☐ Short-field takeoff and maximum performance climb - FM 208-214; POH-4
☐☐☐ Short-field approach and landing - FM 215-221; POH-4
☐☐☐ Go-around/rejected landing - FM 231-235; POH-4
☐☐☐ Collision and runway incursion avoidance procedures - FM 153-154
☐☐☐ Straight-and-level (IR) - CFI
☐☐☐ Constant airspeed climbs and descents (IR) - CFI
☐☐☐ Turns to a heading (IR) - CFI
☐☐☐ Recovery from unusual attitudes (IR) - CFI
☐☐☐ Partial panel flying (IR) - CFI
☐☐☐ Intercepting and tracking navigational systems (IR) - CFI
☐☐☐ After landing, parking, and securing - FM 336-339; POH-4

________________________________________

4. New items (night operations)

☐☐☐ Weather information - FM 77-91
☐☐☐ Cross-country flight planning - FM 92-94
☐☐☐ Pilotage and dead reckoning - FM 270-274
☐☐☐ Navigation systems and radar services - FM 275-276; PH 481-514
☐☐☐ Magnetic compass turns - FM 272-274; PH 96-101
☐☐☐ Diversion - FM 277-279
☐☐☐ Lost procedures - FM 280-282
☐☐☐ Emergency equipment and survival gear - FM 323-324; POH-7, 9
☐☐☐ Additional items at CFI's discretion ________________________________________

5. Postflight critique and preview of next lesson

NOTE: This lesson may be completed under VFR or IFR at the discretion of the flight instructor and/or chief flight instructor, as appropriate.

### Completion Standards

This lesson will have been successfully completed when the pilot is able to plan and conduct a nighttime cross-country flight. The pilot will be able to maintain altitude, ±100 ft.; heading, ±10°; and the desired airspeed, ±10 kt. Additionally, the pilot will be able to verify the airplane's position within 2 NM of the flight planned route at all times.

**Instructor's comments:** ________________________________________

**Lesson assignment:** ________________________________________

**Notes:** ________________________________________

________________________________________

## FLIGHT LESSON 22: SOLO NIGHT--LOCAL

### Objective

To increase the pilot's confidence and proficiency in conducting a night flight. Parts 141 and 61 require 5 hr. in nighttime conditions, with 10 takeoffs and landings (each landing involving a flight in the traffic pattern) at an airport with an operating control tower. This requirement must be met during the three solo night flights--Flight Lessons 22, 23, and 28.

### Text References

*Commercial Pilot Flight Maneuvers and Practical Test Prep* (FM)
*Pilot Handbook* (PH)
Pilot's Operating Handbook (POH)

### Content

1. Flight Lesson 21 complete? Yes ____ Copy of lesson placed in pilot's folder? Yes ____
2. Preflight briefing
3. Review items
   - ☐☐☐ Physiological aspects of night flying - CFI
   - ☐☐☐ Lighting and equipment for night flight - CFI
   - ☐☐☐ Normal and crosswind takeoff and climb - FM 170-178; POH-4
   - ☐☐☐ Normal and crosswind approach and landing - FM 179-196; POH-4
   - ☐☐☐ Short-field takeoff and maximum performance climb - FM 208-214; POH-4
   - ☐☐☐ Short-field approach and landing - FM 215-221; POH-4
   - ☐☐☐ Maneuvering during slow flight - FM 284-288
   - ☐☐☐ Traffic patterns - FM 153-159
   - ☐☐☐ Additional items at CFI's discretion ____________________

     ____________________
4. Postflight critique and preview of next lesson

### Completion Standards

The lesson will have been successfully completed when the pilot completes the solo night flight. The pilot will gain confidence and proficiency in night-flying operations, including takeoffs and landings at an airport with an operating control tower.

**Instructor's comments:** ____________________

____________________

**Lesson assignment:** ____________________

____________________

**Notes:** ____________________

____________________

____________________

____________________

____________________

____________________

____________________

____________________

____________________

____________________

____________________

## FLIGHT LESSON 23: SOLO NIGHT CROSS-COUNTRY

### Objective

To increase the pilot's proficiency in the conduct of nighttime cross-country flights. This solo cross-country flight must include a landing at an airport more than 50 NM from the original point of departure.

### Text References

*Commercial Pilot Flight Maneuvers and Practical Test Prep* (FM)
*Pilot Handbook* (PH)
Pilot's Operating Handbook (POH)

### Content

1. Flight Lesson 22 complete? Yes ____ Copy of lesson placed in pilot's folder? Yes ____
2. Preflight briefing
   - ☐☐☐ Instructor review of pilot's cross-country planning - CFI
3. Review items
   - ☐☐☐ Physiological aspects of night flying - CFI
   - ☐☐☐ Lighting and equipment for night flight - CFI
   - ☐☐☐ Weather information - FM 77-91
   - ☐☐☐ Cross-country flight planning - FM 92-94
   - ☐☐☐ Performance and limitations - FM 102-105; POH-2, 5, 6
   - ☐☐☐ Soft-field takeoff and climb - FM 197-202; POH-4
   - ☐☐☐ Soft-field approach and landing - FM 203-207; POH-4
   - ☐☐☐ Short-field takeoff and maximum performance climb - FM 208-214; POH-4
   - ☐☐☐ Short-field approach and landing - FM 215-221; POH-4
   - ☐☐☐ Pilotage and dead reckoning - FM 270-274
   - ☐☐☐ Navigation systems and radar services - FM 275-276; PH 481-514
   - ☐☐☐ Collision and runway incursion avoidance procedures - FM 153-154
   - ☐☐☐ Traffic patterns - FM 153-159
   - ☐☐☐ Additional items at CFI's discretion ____________________

     ____________________
4. Postflight critique and preview of next lesson

### Completion Standards

The lesson will have been successfully completed when the pilot can plan and conduct the solo nighttime cross-country flight using pilotage, dead reckoning, and navigation systems. During the postflight critique, the instructor will determine how well the flight was conducted through oral questioning.

**Instructor's comments:** ____________________

____________________

**Lesson assignment:** ____________________

____________________

**Notes:** ____________________

____________________

____________________

____________________

____________________

____________________

____________________

____________________

## FLIGHT LESSON 24: SOLO CROSS-COUNTRY

### Objective

To increase the pilot's proficiency in cross-country flights. This flight must include a landing at an airport more than 50 NM from the original point of departure.

### Text References

*Commercial Pilot Flight Maneuvers and Practical Test Prep* (FM)
*Pilot Handbook* (PH)
Pilot's Operating Handbook (POH)

### Content

1. Flight Lesson 23 complete? Yes ____ Copy of lesson placed in pilot's folder? Yes ____
2. Preflight briefing
3. Review items
   - ☐☐☐ Weather information - FM 77-91
   - ☐☐☐ Cross-country flight planning - FM 92-94
   - ☐☐☐ Performance and limitations - FM 102-105; POH-2, 5, 6
   - ☐☐☐ Preflight procedures - FM 118-147; POH-4
   - ☐☐☐ Airport operations - FM 149-167
   - ☐☐☐ Soft-field takeoff and climb - FM 197-202; POH-4
   - ☐☐☐ Soft-field approach and landing - FM 203-207; POH-4
   - ☐☐☐ Short-field takeoff and maximum performance climb - FM 208-214; POH-4
   - ☐☐☐ Short-field approach and landing - FM 215-221; POH-4
   - ☐☐☐ Pilotage and dead reckoning - FM 270-274
   - ☐☐☐ Navigation systems and radar services - FM 275-276; PH 481-514
   - ☐☐☐ Postflight procedures - FM 335-339
   - ☐☐☐ Additional items at CFI's discretion ________________________________________

     ________________________________________
4. Postflight critique and preview of next lesson

### Completion Standards

The lesson will have been successfully completed when the pilot completes the cross-country flight. The pilot's proficiency will increase to meet the skill level of a commercial pilot.

**Instructor's comments:** ________________________________________

________________________________________

**Lesson assignment:** ________________________________________

________________________________________

**Notes:** ________________________________________

________________________________________

________________________________________

________________________________________

________________________________________

________________________________________

________________________________________

________________________________________

________________________________________

________________________________________

________________________________________

## FLIGHT LESSON 25: INSTRUMENT REVIEW

### Objective

To review and increase pilot proficiency in the basic instrument maneuvers.

### Text References

*Commercial Pilot Flight Maneuvers and Practical Test Prep* (FM)
*Pilot Handbook* (PH)

### Content

1. Flight Lesson 24 complete? Yes ____ Copy of lesson placed in pilot's folder? Yes ____
2. Preflight briefing
3. Review items (all to be completed under simulated instrument conditions)
   - ☐☐☐ Straight-and-level - CFI
   - ☐☐☐ Constant airspeed climbs - CFI
   - ☐☐☐ Constant airspeed descents - CFI
   - ☐☐☐ Turns to a heading - CFI
   - ☐☐☐ Partial panel flying - CFI
   - ☐☐☐ Maneuvering during slow flight - FM 284-288
   - ☐☐☐ Power-off stalls - FM 289-294; PH 71-75
   - ☐☐☐ Power-on stalls - FM 295-300; PH 71-75
   - ☐☐☐ Recovery from unusual flight attitudes - CFI
   - ☐☐☐ Intercepting and tracking navigation systems - CFI
   - ☐☐☐ Navigation systems and radar services - FM 275-276; PH 481-514
   - ☐☐☐ Additional items at CFI's discretion ____________________
4. Postflight critique and preview of next lesson

### Completion Standards

This lesson will have been successfully completed when the pilot performs the maneuvers at an instrument pilot skill level. The pilot will be able to maintain the desired altitude, ±100 ft.; airspeed, ±10 kt.; and heading, ±10°, and to track a desired radial.

**Instructor's comments:** ____________________

____________________

**Lesson assignment:** ____________________

____________________

**Notes:** ____________________

## FLIGHT LESSON 26: SOLO CROSS-COUNTRY

### Objective

To increase the pilot's proficiency in cross-country flights. This flight must include a landing at an airport more than 50 NM from the original point of departure.

### Text References

*Commercial Pilot Flight Maneuvers and Practical Test Prep* (FM)
*Pilot Handbook* (PH)
Pilot's Operating Handbook (POH)

### Content

1. Flight Lesson 25 complete? Yes ____ Copy of lesson placed in pilot's folder? Yes ____
2. Preflight briefing
3. Review items
   - ☐☐☐ Weather information - FM 77-91
   - ☐☐☐ Cross-country flight planning - FM 92-94
   - ☐☐☐ Performance and limitations - FM 102-105; POH-2, 5, 6
   - ☐☐☐ Preflight procedures - FM 118-147; POH-4
   - ☐☐☐ Airport operations - FM 149-167
   - ☐☐☐ Soft-field takeoff and climb - FM 197-202; POH-4
   - ☐☐☐ Soft-field approach and landing - FM 203-207; POH-4
   - ☐☐☐ Short-field takeoff and maximum performance climb - FM 208-214; POH-4
   - ☐☐☐ Short-field approach and landing - FM 215-221; POH-4
   - ☐☐☐ Pilotage and dead reckoning - FM 270-274
   - ☐☐☐ Navigation systems and radar services - FM 275-276; PH 481-514
   - ☐☐☐ Postflight procedures - FM 335-339
   - ☐☐☐ Additional items at CFI's discretion ____________________

     ____________________
4. Postflight critique and preview of next lesson

### Completion Standards

The lesson will have been successfully completed when the pilot completes the cross-country flight. The pilot's proficiency will increase to meet the skill level of a commercial pilot.

**Instructor's comments:** ____________________

____________________

**Lesson assignment:** ____________________

____________________

**Notes:** ____________________

____________________

____________________

____________________

____________________

____________________

____________________

____________________

____________________

____________________

____________________

## FLIGHT LESSON 27: SOLO CROSS-COUNTRY

### Objective

To increase the pilot's proficiency in cross-country flights. This flight must include a landing at an airport more than 50 NM from the original point of departure.

### Text References

*Commercial Pilot Flight Maneuvers and Practical Test Prep* (FM)
*Pilot Handbook* (PH)
Pilot's Operating Handbook (POH)

### Content

1. Flight Lesson 26 complete? Yes ____ Copy of lesson placed in pilot's folder? Yes ____
2. Preflight briefing
3. Review items
   - ☐☐☐ Weather information - FM 77-91
   - ☐☐☐ Cross-country flight planning - FM 92-94
   - ☐☐☐ Performance and limitations - FM 102-105; POH-2, 5, 6
   - ☐☐☐ Preflight procedures - FM 118-147; POH-4
   - ☐☐☐ Airport operations - FM 149-167
   - ☐☐☐ Soft-field takeoff and climb - FM 197-202; POH-4
   - ☐☐☐ Soft-field approach and landing - FM 203-207; POH-4
   - ☐☐☐ Short-field takeoff and maximum performance climb - FM 208-214; POH-4
   - ☐☐☐ Short-field approach and landing - FM 215-221; POH-4
   - ☐☐☐ Pilotage and dead reckoning - FM 270-274
   - ☐☐☐ Navigation systems and radar services - FM 275-276; PH 481-514
   - ☐☐☐ Postflight procedures - FM 335-339
   - ☐☐☐ Additional items at CFI's discretion ____________________

     ____________________
4. Postflight critique and preview of next lesson

### Completion Standards

The lesson will have been successfully completed when the pilot completes the cross-country flight. The pilot's proficiency will increase to meet the skill level of a commercial pilot.

**Instructor's comments:** ____________________

____________________

**Lesson assignment:** ____________________

____________________

**Notes:** ____________________

____________________

____________________

____________________

____________________

____________________

____________________

____________________

____________________

____________________

____________________

## FLIGHT LESSON 28: SOLO NIGHT--LOCAL

### Objective

To provide the pilot with experience with night operations and increase the pilot's proficiency when operating in these conditions.

### Text References

*Commercial Pilot Flight Maneuvers and Practical Test Prep* (FM)
*Pilot Handbook* (PH)
Pilot's Operating Handbook (POH)

### Content

1. Flight Lesson 27 complete? Yes ___ Copy of lesson placed in pilot's folder? Yes ___
2. Preflight briefing
3. Review items
   - ☐☐☐ Weather information - FM 77-91
   - ☐☐☐ Aeromedical factors - FM 111-116
   - ☐☐☐ Physiological aspects of night flying - CFI
   - ☐☐☐ Lighting and equipment for night flight - CFI
   - ☐☐☐ Traffic patterns - FM 153-159
   - ☐☐☐ Normal and crosswind takeoff and climb - FM 170-178; POH-4
   - ☐☐☐ Normal and crosswind approach and landing - FM 179-196; POH-4
   - ☐☐☐ Short-field takeoff and maximum performance climb - FM 208-214; POH-4
   - ☐☐☐ Short-field approach and landing - FM 215-221; POH-4
   - ☐☐☐ Power-off stalls - FM 289-294; PH 71-75
   - ☐☐☐ Power-on stalls - FM 295-300; PH 71-75
   - ☐☐☐ Additional items at CFI's discretion ___________________________

     ___________________________
4. Postflight critique and preview of next lesson

### Completion Standards

The lesson will have been successfully completed when the pilot completes the listed maneuvers assigned for this solo night flight. With this flight lesson being the last solo night flight, the pilot is required to have at least 5 hr. of solo flight time in nighttime conditions with 10 takeoffs and landings (each involving a flight with a traffic pattern) at an airport with an operating control tower.

**Instructor's comments:** ___________________________

___________________________

**Lesson assignment:** ___________________________

___________________________

**Notes:** ___________________________

___________________________

___________________________

___________________________

___________________________

___________________________

___________________________

___________________________

___________________________

___________________________

## FLIGHT LESSON 29: SOLO CROSS-COUNTRY

### Objective

To increase the pilot's proficiency in cross-country flights. This flight must include a landing at an airport more than 50 NM from the original point of departure.

### Text References

*Commercial Pilot Flight Maneuvers and Practical Test Prep* (FM)
*Pilot Handbook* (PH)
Pilot's Operating Handbook (POH)

### Content

1. Flight Lesson 28 complete? Yes ____ Copy of lesson placed in pilot's folder? Yes ____
2. Preflight briefing
3. Review items
   - ☐☐☐ Weather information - FM 77-91
   - ☐☐☐ Cross-country flight planning - FM 92-94
   - ☐☐☐ Performance and limitations - FM 102-105; POH-2, 5, 6
   - ☐☐☐ Preflight procedures - FM 118-147; POH-4
   - ☐☐☐ Airport operations - FM 149-167
   - ☐☐☐ Soft-field takeoff and climb - FM 197-202; POH-4
   - ☐☐☐ Soft-field approach and landing - FM 203-207; POH-4
   - ☐☐☐ Short-field takeoff and maximum performance climb - FM 208-214; POH-4
   - ☐☐☐ Short-field approach and landing - FM 215-221; POH-4
   - ☐☐☐ Pilotage and dead reckoning - FM 270-274
   - ☐☐☐ Navigation systems and radar services - FM 275-276; PH 481-514
   - ☐☐☐ Postflight procedures - FM 335-339
   - ☐☐☐ Additional items at CFI's discretion ____________________

     ____________________
4. Postflight critique and preview of next lesson

### Completion Standards

The lesson will have been successfully completed when the pilot completes the cross-country flight. The pilot's proficiency will increase to meet the skill level of a commercial pilot.

**Instructor's comments:** ____________________

____________________

**Lesson assignment:** ____________________

____________________

**Notes:** ____________________

____________________

____________________

____________________

____________________

____________________

____________________

____________________

____________________

____________________

____________________

## FLIGHT LESSON 30: SOLO CROSS-COUNTRY

### Objective

To increase the pilot's proficiency in cross-country flights. This flight must include a landing at an airport more than 50 NM from the original point of departure.

### Text References

*Commercial Pilot Flight Maneuvers and Practical Test Prep* (FM)
*Pilot Handbook* (PH)
Pilot's Operating Handbook (POH)

### Content

1. Flight Lesson 29 complete? Yes ____ Copy of lesson placed in pilot's folder? Yes ____
2. Preflight briefing
3. Review items
   - ☐☐☐ Weather information - FM 77-91
   - ☐☐☐ Cross-country flight planning - FM 92-94
   - ☐☐☐ Performance and limitations - FM 102-105; POH-2, 5, 6
   - ☐☐☐ Preflight procedures - FM 118-147; POH-4
   - ☐☐☐ Airport operations - FM 149-167
   - ☐☐☐ Soft-field takeoff and climb - FM 197-202; POH-4
   - ☐☐☐ Soft-field approach and landing - FM 203-207; POH-4
   - ☐☐☐ Short-field takeoff and maximum performance climb - FM 208-214; POH-4
   - ☐☐☐ Short-field approach and landing - FM 215-221; POH-4
   - ☐☐☐ Pilotage and dead reckoning - FM 270-274
   - ☐☐☐ Navigation systems and radar services - FM 275-276; PH 481-514
   - ☐☐☐ Postflight procedures - FM 335-339
   - ☐☐☐ Additional items at CFI's discretion ____________________

     ____________________
4. Postflight critique and preview of next lesson

### Completion Standards

The lesson will have been successfully completed when the pilot completes the cross-country flight. The pilot's proficiency will increase to meet the skill level of a commercial pilot.

**Instructor's comments:** ____________________

____________________

**Lesson assignment:** ____________________

____________________

**Notes:** ____________________

____________________

____________________

____________________

____________________

____________________

____________________

____________________

____________________

____________________

____________________

## FLIGHT LESSON 31: MANEUVERS REVIEW

### Objective

To review procedures and maneuvers covered previously. Additionally, the pilot will plan and depart on a VFR cross-country flight so the instructor can evaluate the pilot's proficiency with basic navigation, lost procedures, and diversion to an alternate airport.

### Text References

*Commercial Pilot Flight Maneuvers and Practical Test Prep* (FM)
Pilot's Operating Handbook (POH)

### Content

1. Flight Lesson 30 complete? Yes ____ Copy of lesson placed in pilot's folder? Yes ____
2. Preflight briefing
3. Review items
   - ☐☐☐ Weather information - FM 77-91
   - ☐☐☐ Cross-country flight planning - FM 92-94
   - ☐☐☐ Performance and limitations - FM 102-105; POH-2, 5, 6
   - ☐☐☐ Operation of systems - FM 106-110; POH-1, 7, 8, 9
   - ☐☐☐ Traffic patterns - FM 153-159
   - ☐☐☐ Normal and crosswind takeoff and climb - FM 170-178; POH-4
   - ☐☐☐ Normal and crosswind approach and landing - FM 179-196; POH-4
   - ☐☐☐ Soft-field takeoff and climb - FM 197-202; POH-4
   - ☐☐☐ Soft-field approach and landing - FM 203-207; POH-4
   - ☐☐☐ Short-field takeoff and maximum performance climb - FM 208-214; POH-4
   - ☐☐☐ Short-field approach and landing - FM 215-221; POH-4
   - ☐☐☐ Power-off 180° accuracy approach and landing - FM 222-230
   - ☐☐☐ Go-around/rejected landing - FM 231-235; POH-4
   - ☐☐☐ Pilotage and dead reckoning - FM 270-274
   - ☐☐☐ Magnetic compass turns - FM 272-274; PH 96-101
   - ☐☐☐ Diversion - FM 277-279
   - ☐☐☐ Lost procedures - FM 280-282
   - ☐☐☐ Straight-and-level (IR) - CFI
   - ☐☐☐ Constant airspeed climbs and descents (IR) - CFI
   - ☐☐☐ Turns to a heading (IR) - CFI
   - ☐☐☐ Recovery from unusual attitudes (IR) - CFI
   - ☐☐☐ Emergency descent - FM 311-313; POH-3
   - ☐☐☐ Emergency approach and landing (simulated) - FM 314-320; POH-3
   - ☐☐☐ Systems and equipment malfunctions - FM 321-322; POH-3, 9
   - ☐☐☐ Emergency equipment and survival gear - FM 323-324; POH-7, 9
   - ☐☐☐ Additional items at CFI's discretion ____________________

     ____________________
4. Postflight critique and preview of next lesson

### Completion Standards

The lesson will have been successfully completed when the pilot demonstrates increased proficiency in planning and flying a VFR cross-country flight. The pilot will be able to maintain the desired altitude, ±100 ft.; airspeed, ±10 kt.; and heading, ±10°, and verify the airplane's position within 2 NM of the flight planned route at all times.

**Instructor's comments:** ____________________

____________________

**Lesson assignment:** ____________________

____________________

**Notes:** ____________________

____________________

## FLIGHT LESSON 32: SOLO CROSS-COUNTRY

### Objective

To increase the pilot's proficiency in cross-country flights. This flight must include a landing at an airport more than 50 NM from the original point of departure.

### Text References

*Commercial Pilot Flight Maneuvers and Practical Test Prep* (FM)
*Pilot Handbook* (PH)
Pilot's Operating Handbook (POH)

### Content

1. Flight Lesson 31 complete? Yes ____ Copy of lesson placed in pilot's folder? Yes ____
2. Preflight briefing
3. Review items
   - ☐☐☐ Weather information - FM 77-91
   - ☐☐☐ Cross-country flight planning - FM 92-94
   - ☐☐☐ Performance and limitations - FM 102-105; POH-2, 5, 6
   - ☐☐☐ Preflight procedures - FM 118-147; POH-4
   - ☐☐☐ Airport operations - FM 149-167
   - ☐☐☐ Soft-field takeoff and climb - FM 197-202; POH-4
   - ☐☐☐ Soft-field approach and landing - FM 203-207; POH-4
   - ☐☐☐ Short-field takeoff and maximum performance climb - FM 208-214; POH-4
   - ☐☐☐ Short-field approach and landing - FM 215-221; POH-4
   - ☐☐☐ Pilotage and dead reckoning - FM 270-274
   - ☐☐☐ Navigation systems and radar services - FM 275-276; PH 481-514
   - ☐☐☐ Postflight procedures - FM 335-339
   - ☐☐☐ Additional items at CFI's discretion ____________________
4. Postflight critique and preview of next lesson

### Completion Standards

The lesson will have been successfully completed when the pilot completes the cross-country flight. The pilot's proficiency will increase to meet the skill level of a commercial pilot.

**Instructor's comments:** ____________________

**Lesson assignment:** ____________________

**Notes:** ____________________

## FLIGHT LESSON 33: SOLO CROSS-COUNTRY

### Objective

To increase the pilot's proficiency in cross-country flights. In accordance with Part 61, this flight must be at least 300 NM total distance. For both Part 61 and Part 141 training, the flight must include landings at a minimum of three points and one segment of the flight consisting of a straight-line distance of at least 250 NM. In Hawaii, one segment must have a straight-line distance of 150 NM.

### Text References

*Commercial Pilot Flight Maneuvers and Practical Test Prep* (FM)
*Pilot Handbook* (PH)
Pilot's Operating Handbook (POH)

### Content

1. Flight Lesson 32 complete? Yes ____ Copy of lesson placed in pilot's folder? Yes ____
2. Preflight briefing
3. Review items
   - ☐☐☐ Weather information - FM 77-91
   - ☐☐☐ Cross-country flight planning - FM 92-94
   - ☐☐☐ Performance and limitations - FM 102-105; POH-2, 5, 6
   - ☐☐☐ Preflight procedures - FM 118-147; POH-4
   - ☐☐☐ Soft-field takeoff and climb - FM 197-202; POH-4
   - ☐☐☐ Soft-field approach and landing - FM 203-207; POH-4
   - ☐☐☐ Short-field takeoff and maximum performance climb - FM 208-214; POH-4
   - ☐☐☐ Short-field approach and landing - FM 215-221; POH-4
   - ☐☐☐ Pilotage and dead reckoning - FM 270-274
   - ☐☐☐ Navigation systems and radar services - FM 275-276; PH 481-514
   - ☐☐☐ Postflight procedures - FM 335-339
   - ☐☐☐ Additional items at CFI's discretion ____________________

     ____________________
4. Postflight critique and preview of next lesson

### Completion Standards

The lesson will have been successfully completed when the pilot completes the cross-country flight. During the preflight briefing, the pilot will show accurate flight planning and display good judgment in making a go/no-go decision.

**Instructor's comments:** ____________________

____________________

**Lesson assignment:** ____________________

____________________

**Notes:** ____________________

____________________

____________________

____________________

____________________

____________________

____________________

____________________

____________________

## FLIGHT LESSON 34: STAGE TWO CHECK

### Objective

During this stage check, an authorized instructor will determine if the pilot is proficient in planning and conducting daytime and nighttime cross-country flights at the commercial pilot skill level.

### Text References

*Commercial Pilot Flight Maneuvers and Practical Test Prep* (FM)
*Pilot Handbook* (PH)
Pilot's Operating Handbook (POH)

### Content

1. Flight Lesson 33 complete? Yes ____ Copy of lesson placed in pilot's folder? Yes ____
2. Preflight briefing
3. Stage check tasks
   - ☐☐☐ Weather information - FM 77-91
   - ☐☐☐ Cross-country flight planning - FM 92-94
   - ☐☐☐ National Airspace System - FM 95-101
   - ☐☐☐ Performance and limitations - FM 102-105; POH-2, 5, 6
   - ☐☐☐ Aeronautical decision making (ADM) - PH 343-349
   - ☐☐☐ Cockpit management - FM 123-126
   - ☐☐☐ Radio communications and ATC light signals - FM 150-152
   - ☐☐☐ Runway incursion avoidance - FM 139-143
   - ☐☐☐ Traffic patterns - FM 153-159
   - ☐☐☐ Airport, runway, and taxiway signs, markings, and lighting - FM 160-167
   - ☐☐☐ Soft-field takeoff and climb - FM 197-202; POH-4
   - ☐☐☐ Soft-field approach and landing - FM 203-207; POH-4
   - ☐☐☐ Short-field takeoff and maximum performance climb - FM 208-214; POH-4
   - ☐☐☐ Short-field approach and landing - FM 215-221; POH-4
   - ☐☐☐ Pilotage and dead reckoning - FM 270-274
   - ☐☐☐ Magnetic compass turns - FM 272-274; PH 96-101
   - ☐☐☐ Navigation systems and radar services - FM 275-276; PH 481-514
   - ☐☐☐ Diversion - FM 277-279
   - ☐☐☐ Lost procedures - FM 280-282
   - ☐☐☐ Physiological aspects of night flying - CFI
   - ☐☐☐ Lighting and equipment for night flight - CFI
   - ☐☐☐ Systems and equipment malfunctions - FM 321-322; POH-3, 9
   - ☐☐☐ Postflight procedures - FM 335-339
   - ☐☐☐ Additional items at CFI's discretion ____________________

     ____________________
4. Postflight critique and preview of next lesson

### Completion Standards

The lesson and Stage Two will have been successfully completed when the pilot can demonstrate proficiency in cross-country planning and flight procedures in daytime and nighttime conditions. The flight does not have to take place at night. The pilot may demonstrate understanding of night flight considerations in lieu of conducting the flight at night. The pilot will be able to maintain the desired altitude, ±100 ft.; airspeed, ±10 kt.; and heading, ±10°, and verify the airplane's position within 2 NM of the flight planned route at all times.

**Instructor's comments:** ____________________

____________________

**Lesson assignment:** ____________________

____________________

**Notes:** ____________________

____________________

## STAGE THREE

### Stage Three Objective

To provide the pilot with the instruction and practice necessary to attain the skill level required to pass the commercial pilot airplane (single-engine land) practical test.

### Stage Three Completion Standards

The stage will be completed when the pilot demonstrates proficiency in all of the tasks of the commercial pilot airplane (single-engine land) practical test and meets or exceeds the minimum acceptable standards for the commercial pilot certificate.

| Lesson | Topic |
|---|---|
| 35 | Complex Airplane--Performance Maneuvers |
| 36 | Complex Airplane--Eights-on-Pylons |
| 37 | Solo Practice |
| 38 | Solo Practice |
| 39 | Maneuvers Review |
| 40 | Maneuvers Review |
| 41 | Solo Practice |
| 42 | Solo Practice |
| 43 | Complex Airplane--Maneuvers Review |
| 44 | Solo Practice |
| 45 | Maneuvers Review |
| 46 | Solo Practice |
| 47 | Maneuvers Review |
| 48 | Maneuvers Review |
| 49 | Solo Practice |
| 50 | Solo Practice |
| 51 | Complex Airplane--Maneuvers Review |
| 52 | Solo Practice |
| 53 | Maneuvers Review |
| 54 | Dual Cross-Country |
| 55 | Solo Practice |
| 56 | Maneuvers Review |
| 57 | Solo Practice |
| 58 | Complex Airplane--Maneuvers Review |
| 59 | Stage Three Check--Complex Airplane |
| 60 | End-of-Course Check--Complex Airplane |

## FLIGHT LESSON 35: COMPLEX AIRPLANE--PERFORMANCE MANEUVERS

### Objective

To review procedures and maneuvers covered previously. Additionally, the pilot will be introduced to performance maneuvers in a complex airplane.

### Text References

*Commercial Pilot Flight Maneuvers and Practical Test Prep* (FM)
Pilot's Operating Handbook (POH)

### Content

1. Flight Lesson 34 complete? Yes ____ Copy of lesson placed in pilot's folder? Yes ____
2. Preflight briefing
3. Review items
   - ☐☐☐ Operation of systems - FM 106-110; POH-1, 7, 8, 9
   - ☐☐☐ Aeronautical decision making (ADM) - PH 343-349
   - ☐☐☐ Soft-field takeoff and climb - FM 197-202; POH-4
   - ☐☐☐ Soft-field approach and landing - FM 203-207; POH-4
   - ☐☐☐ Traffic patterns - FM 153-159
   - ☐☐☐ Go-around/rejected landing - FM 231-235; POH-4

   ______________________________
4. New items
   - ☐☐☐ Steep turns - FM 238-242
   - ☐☐☐ Steep spirals - FM 243-248
   - ☐☐☐ Chandelles - FM 249-253
   - ☐☐☐ Lazy eights - FM 254-259
   - ☐☐☐ Additional items at CFI's discretion ______________________________

   ______________________________
5. Postflight critique and preview of next lesson

### Completion Standards

The lesson will have been successfully completed when the pilot demonstrates an understanding of performing steep turns, steep spirals, chandelles, and lazy eights in the complex airplane. The pilot will be able to maintain $V_Y$, ±5 kt. during the takeoff climb and maintain the recommended approach airspeed (with gust factors applied), ±5 kt.

**Instructor's comments:** ______________________________

______________________________

**Lesson assignment:** ______________________________

______________________________

**Notes:** ______________________________

______________________________

______________________________

______________________________

______________________________

______________________________

______________________________

______________________________

______________________________

______________________________

## FLIGHT LESSON 36: COMPLEX AIRPLANE--EIGHTS-ON-PYLONS

### Objective

To review procedures and maneuvers covered previously. Additionally, the pilot will be introduced to eights-on-pylons and basic attitude instrument flight in a complex airplane.

### Text References

*Commercial Pilot Flight Maneuvers and Practical Test Prep* (FM)
Pilot's Operating Handbook (POH)

### Content

1. Flight Lesson 35 complete? Yes ____ Copy of lesson placed in pilot's folder? Yes ____
2. Preflight briefing
3. Review items
   - ☐☐☐ Operation of systems - FM 106-110; POH-1, 7, 8, 9
   - ☐☐☐ Supplemental oxygen - FM 326-327; POH-7, 8
   - ☐☐☐ Pressurization - FM 328-334; POH-7
   - ☐☐☐ Aeromedical factors - FM 111-116
   - ☐☐☐ Short-field takeoff and maximum performance climb - FM 208-214; POH-4
   - ☐☐☐ Short-field approach and landing - FM 215-221; POH-4
   - ☐☐☐ Power-off 180° accuracy approach and landing - FM 222-230
   - ☐☐☐ Go-around/rejected landing - FM 231-235; POH-4
   - ☐☐☐ Chandelles - FM 249-253
   - ☐☐☐ Lazy eights - FM 254-259
   - ☐☐☐ Systems and equipment malfunctions - FM 321-322; POH-3, 9

   ______________________________________________
4. New items
   - ☐☐☐ Eights-on-pylons - FM 262-267
   - ☐☐☐ Straight-and-level (IR) - CFI
   - ☐☐☐ Turns to a heading (IR) - CFI
   - ☐☐☐ Recovery from unusual flight attitudes (IR) - CFI
   - ☐☐☐ Constant airspeed climbs and descents (IR) - CFI
   - ☐☐☐ Additional items at CFI's discretion ______________________________________________

   ______________________________________________
5. Postflight critique and preview of next lesson

### Completion Standards

The lesson will have been successfully completed when the pilot demonstrates an understanding of performing eights-on-pylons in the complex airplane. Additionally, the pilot's proficiency in chandelles and lazy eights will increase. The pilot will be able to maintain $V_X$, +10/–0 kt. until clearing the obstacles during the short-field takeoff, then maintain $V_Y$, ±5 kt. During the short-field and power-off accuracy approaches and landings, the pilot will be able to maintain the recommended approach airspeed (with gust factors applied), ±5 kt., and touch down at or within 200 ft. beyond a specific point.

**Instructor's comments:** ______________________________________________

______________________________________________

**Lesson assignment:** ______________________________________________

______________________________________________

**Notes:** ______________________________________________

______________________________________________

______________________________________________

______________________________________________

______________________________________________

## FLIGHT LESSON 37: SOLO PRACTICE

### Objective

To further develop the pilot's proficiency through solo practice of assigned maneuvers.

### Text References

*Commercial Pilot Flight Maneuvers and Practical Test Prep* (FM)
Pilot's Operating Handbook (POH)

### Content

1. Flight Lesson 36 complete? Yes ____ Copy of lesson placed in pilot's folder? Yes ____
2. Preflight briefing
3. Review items
   - ☐☐☐ Soft-field takeoff and climb - FM 197-202; POH-4
   - ☐☐☐ Soft-field approach and landing - FM 203-207; POH-4
   - ☐☐☐ Short-field takeoff and maximum performance climb - FM 208-214; POH-4
   - ☐☐☐ Short-field approach and landing - FM 215-221; POH-4
   - ☐☐☐ Power-off 180° accuracy approach and landing - FM 222-230
   - ☐☐☐ Steep turns - FM 238-242
   - ☐☐☐ Steep spirals - FM 243-248
   - ☐☐☐ Chandelles - FM 249-253
   - ☐☐☐ Lazy eights - FM 254-259
   - ☐☐☐ Eights-on-pylons - FM 262-267
   - ☐☐☐ Additional items at CFI's discretion ____________________
4. Postflight critique and preview of next lesson

### Completion Standards

The lesson will have been successfully completed when the pilot completes each maneuver assigned for this solo flight. The pilot will gain proficiency as a result of the solo flight.

**Instructor's comments:** ____________________

**Lesson assignment:** ____________________

**Notes:** ____________________

## FLIGHT LESSON 38: SOLO PRACTICE

### Objective

To further develop the pilot's proficiency through solo practice of the assigned maneuvers.

### Text References

*Commercial Pilot Flight Maneuvers and Practical Test Prep* (FM)
*Pilot Handbook* (PH)
Pilot's Operating Handbook (POH)

### Content

1. Flight Lesson 37 complete? Yes ____ Copy of lesson placed in pilot's folder? Yes ____
2. Preflight briefing
3. Review items
   - ☐☐☐ Soft-field takeoff and climb - FM 197-202; POH-4
   - ☐☐☐ Soft-field approach and landing - FM 203-207; POH-4
   - ☐☐☐ Traffic patterns - FM 153-159
   - ☐☐☐ Steep turns - FM 238-242
   - ☐☐☐ Steep spirals - FM 243-248
   - ☐☐☐ Chandelles - FM 249-253
   - ☐☐☐ Lazy eights - FM 254-259
   - ☐☐☐ Eights-on-pylons - FM 262-267
   - ☐☐☐ Additional items at CFI's discretion ____________________

     ____________________
4. Postflight critique and preview of next lesson

### Completion Standards

The lesson will have been successfully completed when the pilot completes each maneuver assigned for this solo flight. The pilot will gain proficiency as a result of the solo flight.

**Instructor's comments:** ____________________

____________________

**Lesson assignment:** ____________________

____________________

**Notes:** ____________________

____________________

____________________

____________________

____________________

____________________

____________________

____________________

____________________

____________________

____________________

____________________

____________________

____________________

## FLIGHT LESSON 39: MANEUVERS REVIEW

### Objective

To review procedures and maneuvers covered previously.

### Text References

*Commercial Pilot Flight Maneuvers and Practical Test Prep* (FM)
*Pilot Handbook* (PH)
Pilot's Operating Handbook (POH)

### Content

1. Flight Lesson 38 complete? Yes ____ Copy of lesson placed in pilot's folder? Yes ____
2. Preflight briefing
3. Review items
   - ☐☐☐ Radio communications and ATC light signals - FM 150-152
   - ☐☐☐ Airport, runway, and taxiway signs, markings, and lighting - FM 160-167
   - ☐☐☐ Soft-field takeoff and climb - FM 197-202; POH-4
   - ☐☐☐ Soft-field approach and landing - FM 203-207; POH-4
   - ☐☐☐ Traffic patterns - FM 153-159
     - ☐☐☐ Runway incursion and collision avoidance - FM 153-154
     - ☐☐☐ Wake turbulence avoidance - FM 154-155
     - ☐☐☐ Wind shear avoidance - FM 155
   - ☐☐☐ Maneuvering during slow flight - FM 284-288
   - ☐☐☐ Power-off stalls - FM 289-294; PH 71-75
   - ☐☐☐ Power-on stalls - FM 295-300; PH 71-75
   - ☐☐☐ Accelerated stalls - FM 301-305; PH 71-75
   - ☐☐☐ Spin awareness - FM 306-308; POH-2
   - ☐☐☐ Steep turns - FM 238-242
   - ☐☐☐ Steep spirals - FM 243-248
   - ☐☐☐ Chandelles - FM 249-253
   - ☐☐☐ Emergency approach and landing (simulated) - FM 314-320; POH-3
   - ☐☐☐ Additional items at CFI's discretion ____________________

     ____________________
4. Postflight critique and preview of next lesson

### Completion Standards

The lesson will have been successfully completed when the pilot demonstrates increased proficiency in the listed maneuvers. During steep turns, the pilot will be able to maintain the desired altitude, ±150 ft.; airspeed, ±10 kt.; and roll out on the entry heading, ±10°. Additionally, the pilot will be able to perform slow flight and stalls to the minimum standard in the current FAA Commercial Pilot Practical Test Standards.

**Instructor's comments:** ____________________

____________________

**Lesson assignment:** ____________________

____________________

**Notes:** ____________________

____________________

____________________

____________________

____________________

____________________

____________________

## FLIGHT LESSON 40: MANEUVERS REVIEW

### Objective

To review procedures and maneuvers covered previously.

### Text References

*Commercial Pilot Flight Maneuvers and Practical Test Prep* (FM)
Pilot's Operating Handbook (POH)

### Content

1. Flight Lesson 39 complete? Yes ____ Copy of lesson placed in pilot's folder? Yes ____
2. Preflight briefing
3. Review items
   - ☐☐☐ Short-field takeoff and maximum performance climb - FM 208-214; POH-4
   - ☐☐☐ Short-field approach and landing - FM 215-221; POH-4
   - ☐☐☐ Power-off 180° accuracy approach and landing - FM 222-230
   - ☐☐☐ Go-around/rejected landing - FM 231-235; POH-4
   - ☐☐☐ Chandelles - FM 249-253
   - ☐☐☐ Lazy eights - FM 254-259
   - ☐☐☐ Eights-on-pylons - FM 262-267
   - ☐☐☐ Emergency descent - FM 311-313; POH-3
   - ☐☐☐ Emergency approach and landing (simulated) - FM 314-320; POH-3
   - ☐☐☐ Additional items at CFI's discretion ____________________

     ____________________
4. Postflight critique and preview of next lesson

### Completion Standards

The lesson will have been successfully completed when the pilot demonstrates an understanding of the elements of each maneuver, including the correct entry, performance, and recovery techniques. The pilot will be able to perform the short-field takeoffs and landings and the power-off accuracy approach and landing to the minimum standards listed in the current FAA Commercial Pilot Practical Test Standards.

**Instructor's comments:** ____________________

____________________

**Lesson assignment:** ____________________

____________________

**Notes:** ____________________

____________________

____________________

____________________

____________________

____________________

____________________

____________________

____________________

____________________

____________________

____________________

## FLIGHT LESSON 41: SOLO PRACTICE

### Objective

To increase the pilot's proficiency in each of the assigned maneuvers.

### Text References

*Commercial Pilot Flight Maneuvers and Practical Test Prep* (FM)
*Pilot Handbook* (PH)
Pilot's Operating Handbook (POH)

### Content

1. Flight Lesson 40 complete? Yes ____ Copy of lesson placed in pilot's folder? Yes ____
2. Preflight briefing
3. Review items
   - ☐☐☐ Short-field takeoff and maximum performance climb - FM 208-214; POH-4
   - ☐☐☐ Short-field approach and landing - FM 215-221; POH-4
   - ☐☐☐ Power-off 180° accuracy approach and landing - FM 222-230
   - ☐☐☐ Steep turns - FM 238-242
   - ☐☐☐ Steep spirals - FM 243-248
   - ☐☐☐ Chandelles - FM 249-253
   - ☐☐☐ Lazy eights - FM 254-259
   - ☐☐☐ Eights-on-pylons - FM 262-267
   - ☐☐☐ Maneuvering during slow flight - FM 284-288
   - ☐☐☐ Power-off stalls - FM 289-294; PH 71-75
   - ☐☐☐ Emergency descent - FM 311-313; POH-3
   - ☐☐☐ Additional items at CFI's discretion ____________________
4. Postflight critique and preview of next lesson

### Completion Standards

The lesson will have been successfully completed when the pilot completes each maneuver assigned for this solo flight.

**Instructor's comments:** ____________________

**Lesson assignment:** ____________________

**Notes:** ____________________

## FLIGHT LESSON 42: SOLO PRACTICE

### Objective

To increase the pilot's proficiency in each of the assigned maneuvers.

### Text References

*Commercial Pilot Flight Maneuvers and Practical Test Prep* (FM)
*Pilot Handbook* (PH)
Pilot's Operating Handbook (POH)

### Content

1. Flight Lesson 41 complete? Yes ____ Copy of lesson placed in pilot's folder? Yes ____
2. Preflight briefing
3. Review items
   - ☐☐☐ Soft-field takeoff and climb - FM 197-202; POH-4
   - ☐☐☐ Soft-field approach and landing - FM 203-207; POH-4
   - ☐☐☐ Steep turns - FM 238-242
   - ☐☐☐ Steep spirals - FM 243-248
   - ☐☐☐ Chandelles - FM 249-253
   - ☐☐☐ Lazy eights - FM 254-259
   - ☐☐☐ Eights-on-pylons - FM 262-267
   - ☐☐☐ Power-on stalls - FM 295-300; PH 71-75
   - ☐☐☐ Accelerated stalls - FM 301-305; PH 71-75
   - ☐☐☐ Additional items at CFI's discretion ____________________
4. Postflight critique and preview of next lesson

### Completion Standards

The lesson will have been successfully completed when the pilot completes each maneuver assigned for this solo flight.

**Instructor's comments:** ____________________

**Lesson assignment:** ____________________

**Notes:** ____________________

## FLIGHT LESSON 43: COMPLEX AIRPLANE--MANEUVERS REVIEW

### Objective

To review procedures and maneuvers covered previously and to identify areas where additional practice is necessary.

### Text References

*Commercial Pilot Flight Maneuvers and Practical Test Prep* (FM)
*Pilot Handbook* (PH)
Pilot's Operating Handbook (POH)

### Content

1. Flight Lesson 42 complete? Yes ____ Copy of lesson placed in pilot's folder? Yes ____
2. Preflight briefing
3. Review items

☐☐☐ Performance and limitations - FM 102-105; POH-2, 5, 6
☐☐☐ Operation of systems - FM 106-110; POH-1, 7, 8, 9
☐☐☐ Preflight inspection - FM 118-122; POH-4
☐☐☐ Cockpit management - FM 123-126
☐☐☐ Engine starting - FM 127-131; POH-4
☐☐☐ Taxiing - FM 132-138
☐☐☐ Runway incursion avoidance - FM 139-143
☐☐☐ Before-takeoff check - FM 144-147; POH-4
☐☐☐ Traffic patterns - FM 153-159
☐☐☐ Normal and crosswind takeoff and climb - FM 170-178; POH-4
☐☐☐ Normal and crosswind approach and landing - FM 179-196; POH-4
☐☐☐ Soft-field takeoff and climb - FM 197-202; POH-4
☐☐☐ Soft-field approach and landing - FM 203-207; POH-4
☐☐☐ Short-field takeoff and maximum performance climb - FM 208-214; POH-4
☐☐☐ Short-field approach and landing - FM 215-221; POH-4
☐☐☐ Power-off 180° accuracy approach and landing - FM 222-230
☐☐☐ Go-around/rejected landing - FM 231-235; POH-4
☐☐☐ Steep turns - FM 238-242
☐☐☐ Steep spirals - FM 243-248
☐☐☐ Chandelles - FM 249-253
☐☐☐ Lazy eights - FM 254-259
☐☐☐ Eights-on-pylons - FM 262-267
☐☐☐ Maneuvering during slow flight - FM 284-288
☐☐☐ Power-off stalls - FM 289-294; PH 71-75
☐☐☐ Power-on stalls - FM 295-300; PH 71-75
☐☐☐ Accelerated stalls - FM 301-305; PH 71-75
☐☐☐ Spin awareness - FM 306-308; POH-2
☐☐☐ Emergency descent - FM 311-313; POH-3
☐☐☐ Emergency approach and landing (simulated) - FM 314-320; POH-3
☐☐☐ Systems and equipment malfunctions - FM 321-322; POH-3, 9
☐☐☐ Attitude instrument flying (IR) - CFI
☐☐☐ Partial panel flying (IR) - CFI
☐☐☐ Intercepting and tracking navigation systems (IR) - CFI
☐☐☐ Recovery from unusual flight attitudes (IR) - CFI
☐☐☐ After landing, parking, and securing - FM 336-339; POH-4
☐☐☐ Additional items at CFI's discretion ____________

______________________________

______________________________

4. Postflight critique and preview of next lesson

### Completion Standards

The lesson will have been successfully completed when the pilot demonstrates an increased knowledge of the complex airplane systems and flight characteristics. Additionally, the pilot will demonstrate a good understanding of the correct procedures in all of the listed tasks. The pilot will be able to maintain the desired altitude, ±100 ft.; airspeed, ±10 kt.; and heading, ±10°. During takeoffs and landings, the pilot will be able to maintain $V_Y$, ±5 kt.; $V_X$, +5/–0 kt.; and the recommended approach airspeed (with gust factors applied), ±5 kt.

**Instructor's comments:** ______________________________

______________________________

**Lesson assignment:** ______________________________

______________________________

**Notes:** ______________________________

______________________________

______________________________

## FLIGHT LESSON 44: SOLO PRACTICE

### Objective

To increase the pilot's proficiency in each of the assigned maneuvers.

### Text References

*Commercial Pilot Flight Maneuvers and Practical Test Prep* (FM)
*Pilot Handbook* (PH)
Pilot's Operating Handbook (POH)

### Content

1. Flight Lesson 43 complete? Yes ____ Copy of lesson placed in pilot's folder? Yes ____
2. Preflight briefing
3. Review items
   - ☐☐☐ Soft-field takeoff and climb - FM 197-202; POH-4
   - ☐☐☐ Soft-field approach and landing - FM 203-207; POH-4
   - ☐☐☐ Short-field takeoff and maximum performance climb - FM 208-214; POH-4
   - ☐☐☐ Short-field approach and landing - FM 215-221; POH-4
   - ☐☐☐ Power-off 180° accuracy approach and landing - FM 222-230
   - ☐☐☐ Steep turns - FM 238-242
   - ☐☐☐ Steep spirals - FM 243-248
   - ☐☐☐ Chandelles - FM 249-253
   - ☐☐☐ Lazy eights - FM 254-259
   - ☐☐☐ Eights-on-pylons - FM 262-267
   - ☐☐☐ Maneuvering during slow flight - FM 284-288
   - ☐☐☐ Power-off stalls - FM 289-294; PH 71-75
   - ☐☐☐ Power-on stalls - FM 295-300; PH 71-75
   - ☐☐☐ Accelerated stalls - FM 301-305; PH 71-75
   - ☐☐☐ Emergency descent - FM 311-313; POH-3
   - ☐☐☐ Additional items at CFI's discretion ________________________________

     ________________________________
4. Postflight critique and preview of next lesson

### Completion Standards

The lesson will have been successfully completed when the pilot completes each maneuver assigned for this solo flight. The pilot will concentrate on deficient areas from the previous lesson.

**Instructor's comments:** ________________________________

________________________________

**Lesson assignment:** ________________________________

________________________________

**Notes:** ________________________________

________________________________

________________________________

________________________________

________________________________

________________________________

________________________________

________________________________

________________________________

________________________________

## FLIGHT LESSON 45: MANEUVERS REVIEW

### Objective

To review procedures and maneuvers covered previously.

### Text References

*Commercial Pilot Flight Maneuvers and Practical Test Prep* (FM)
Pilot's Operating Handbook (POH)

### Content

1. Flight Lesson 44 complete? Yes ___ Copy of lesson placed in pilot's folder? Yes ___
2. Preflight briefing
3. Review items
   - ☐☐☐ Soft-field takeoff and climb - FM 197-202; POH-4
   - ☐☐☐ Soft-field approach and landing - FM 203-207; POH-4
   - ☐☐☐ Short-field takeoff and maximum performance climb - FM 208-214; POH-4
   - ☐☐☐ Short-field approach and landing - FM 215-221; POH-4
   - ☐☐☐ Power-off 180° accuracy approach and landing - FM 222-230
   - ☐☐☐ Go-around/rejected landing - FM 231-235; POH-4
   - ☐☐☐ Steep turns - FM 238-242
   - ☐☐☐ Steep spirals - FM 243-248
   - ☐☐☐ Chandelles - FM 249-253
   - ☐☐☐ Lazy eights - FM 254-259
   - ☐☐☐ Eights-on-pylons - FM 262-267
   - ☐☐☐ Emergency approach and landing (simulated) - FM 314-320; POH-3
   - ☐☐☐ Additional items at CFI's discretion ____________________
4. Postflight critique and preview of next lesson

### Completion Standards

The lesson will have been successfully completed when the pilot demonstrates an understanding of the correct procedures (entry, performance, and recovery) in each of the assigned maneuvers.

**Instructor's comments:** ____________________

**Lesson assignment:** ____________________

**Notes:** ____________________

## FLIGHT LESSON 46: SOLO PRACTICE

### Objective

To increase the pilot's proficiency in each of the assigned maneuvers.

### Text References

*Commercial Pilot Flight Maneuvers and Practical Test Prep* (FM)
Pilot's Operating Handbook (POH)

### Content

1. Flight Lesson 45 complete? Yes ___ Copy of lesson placed in pilot's folder? Yes ___
2. Preflight briefing
3. Review items
   - ☐☐☐ Soft-field takeoff and climb - FM 197-202; POH-4
   - ☐☐☐ Soft-field approach and landing - FM 203-207; POH-4
   - ☐☐☐ Short-field takeoff and maximum performance climb - FM 208-214; POH-4
   - ☐☐☐ Short-field approach and landing - FM 215-221; POH-4
   - ☐☐☐ Power-off 180° accuracy approach and landing - FM 222-230
   - ☐☐☐ Steep turns - FM 238-242
   - ☐☐☐ Steep spirals - FM 243-248
   - ☐☐☐ Chandelles - FM 249-253
   - ☐☐☐ Lazy eights - FM 254-259
   - ☐☐☐ Eights-on-pylons - FM 262-267
   - ☐☐☐ Additional items at CFI's discretion ______________________________

     ______________________________
4. Postflight critique and preview of next lesson

### Completion Standards

The lesson will have been successfully completed when the pilot completes each maneuver assigned for this solo flight. The pilot will concentrate on deficient areas from the previous lesson.

**Instructor's comments:** ______________________________

______________________________

**Lesson assignment:** ______________________________

______________________________

**Notes:** ______________________________

______________________________

______________________________

______________________________

______________________________

______________________________

______________________________

______________________________

______________________________

______________________________

______________________________

______________________________

______________________________

## FLIGHT LESSON 47: MANEUVERS REVIEW

### Objective

To review procedures and maneuvers covered previously.

### Text References

*Commercial Pilot Flight Maneuvers and Practical Test Prep* (FM)
*Pilot Handbook* (PH)
Pilot's Operating Handbook (POH)

### Content

1. Flight Lesson 46 complete? Yes ____ Copy of lesson placed in pilot's folder? Yes ____
2. Preflight briefing
3. Review items
   - ☐☐☐ Soft-field takeoff and climb - FM 197-202; POH-4
   - ☐☐☐ Soft-field approach and landing - FM 203-207; POH-4
   - ☐☐☐ Short-field takeoff and maximum performance climb - FM 208-214; POH-4
   - ☐☐☐ Short-field approach and landing - FM 215-221; POH-4
   - ☐☐☐ Steep turns - FM 238-242
   - ☐☐☐ Steep spirals - FM 243-248
   - ☐☐☐ Chandelles - FM 249-253
   - ☐☐☐ Lazy eights - FM 254-259
   - ☐☐☐ Eights-on-pylons - FM 262-267
   - ☐☐☐ Maneuvering during slow flight - FM 284-288
   - ☐☐☐ Power-off stalls - FM 289-294; PH 71-75
   - ☐☐☐ Power-on stalls - FM 295-300; PH 71-75
   - ☐☐☐ Accelerated stalls - FM 301-305; PH 71-75
   - ☐☐☐ Additional items at CFI's discretion ____________________

     ____________________
4. Postflight critique and preview of next lesson

### Completion Standards

The lesson will have been successfully completed when the pilot demonstrates an understanding of the elements in each of the assigned maneuvers. The pilot will be able to perform takeoffs, landings, slow flight, and stalls at the minimum standards listed in the current FAA Commercial Pilot Practical Test Standards.

**Instructor's comments:** ____________________

____________________

**Lesson assignment:** ____________________

____________________

**Notes:** ____________________

____________________

____________________

____________________

____________________

____________________

____________________

____________________

____________________

____________________

## FLIGHT LESSON 48: MANEUVERS REVIEW

### Objective

To review procedures and maneuvers covered previously.

### Text References

*Commercial Pilot Flight Maneuvers and Practical Test Prep* (FM)
Pilot's Operating Handbook (POH)

### Content

1. Flight Lesson 47 complete? Yes ___ Copy of lesson placed in pilot's folder? Yes ___
2. Preflight briefing
3. Review items
   - ☐☐☐ Operation of systems - FM 106-110; POH-1, 7, 8, 9
   - ☐☐☐ Soft-field takeoff and climb - FM 197-202; POH-4
   - ☐☐☐ Soft-field approach and landing - FM 203-207; POH-4
   - ☐☐☐ Short-field takeoff and maximum performance climb - FM 208-214; POH-4
   - ☐☐☐ Short-field approach and landing - FM 215-221; POH-4
   - ☐☐☐ Power-off 180° accuracy approach and landing - FM 222-230
   - ☐☐☐ Go-around/rejected landing - FM 231-235; POH-4
   - ☐☐☐ Chandelles - FM 249-253
   - ☐☐☐ Lazy eights - FM 254-259
   - ☐☐☐ Eights-on-pylons - FM 262-267
   - ☐☐☐ Emergency descent - FM 311-313; POH-3
   - ☐☐☐ Emergency approach and landing (simulated) - FM 314-320; POH-3
   - ☐☐☐ Systems and equipment malfunctions - FM 321-322; POH-3, 9
   - ☐☐☐ Emergency equipment and survival gear - FM 323-324; POH-7, 9
   - ☐☐☐ Attitude instrument flying (IR) - CFI
   - ☐☐☐ Recovery from unusual flight attitudes (IR) - CFI
   - ☐☐☐ Partial panel flying (IR) - CFI
   - ☐☐☐ Intercepting and tracking navigation systems (IR) - CFI
   - ☐☐☐ Additional items at CFI's discretion ______________________

     ______________________
4. Postflight critique and preview of next lesson

### Completion Standards

The lesson will have been successfully completed when the pilot demonstrates an increase in proficiency in each of the assigned maneuvers. The pilot will be close to, or meet, the minimum standards listed in the current FAA Commercial Pilot Practical Test Standards.

**Instructor's comments:** ______________________

______________________

**Lesson assignment:** ______________________

______________________

**Notes:** ______________________

______________________

______________________

______________________

______________________

______________________

______________________

______________________

## FLIGHT LESSON 49: SOLO PRACTICE

### Objective

To increase the pilot's proficiency in each of the assigned maneuvers.

### Text References

*Commercial Pilot Flight Maneuvers and Practical Test Prep* (FM)
*Pilot Handbook* (PH)
Pilot's Operating Handbook (POH)

### Content

1. Flight Lesson 48 complete? Yes ____ Copy of lesson placed in pilot's folder? Yes ____
2. Preflight briefing
3. Review items
   - ☐☐☐ Soft-field takeoff and climb - FM 197-202; POH-4
   - ☐☐☐ Soft-field approach and landing - FM 203-207; POH-4
   - ☐☐☐ Short-field takeoff and maximum performance climb - FM 208-214; POH-4
   - ☐☐☐ Short-field approach and landing - FM 215-221; POH-4
   - ☐☐☐ Power-off 180° accuracy approach and landing - FM 222-230
   - ☐☐☐ Steep turns - FM 238-242
   - ☐☐☐ Steep spirals - FM 243-248
   - ☐☐☐ Chandelles - FM 249-253
   - ☐☐☐ Maneuvering during slow flight - FM 284-288
   - ☐☐☐ Power-off stalls - FM 289-294; PH 71-75
   - ☐☐☐ Power-on stalls - FM 295-300; PH 71-75
   - ☐☐☐ Accelerated stalls - FM 301-305; PH 71-75
   - ☐☐☐ Additional items at CFI's discretion ____________________
4. Postflight critique and preview of next lesson

### Completion Standards

The lesson will have been successfully completed when the pilot completes each maneuver assigned for this solo flight.

**Instructor's comments:** ____________________

**Lesson assignment:** ____________________

**Notes:** ____________________

## FLIGHT LESSON 50: SOLO PRACTICE

### Objective

To increase the pilot's proficiency in each of the assigned maneuvers.

### Text References

*Commercial Pilot Flight Maneuvers and Practical Test Prep* (FM)
Pilot's Operating Handbook (POH)

### Content

1. Flight Lesson 49 complete? Yes ___ Copy of lesson placed in pilot's folder? Yes ___
2. Preflight briefing
3. Review items
   - ☐☐☐ Soft-field takeoff and climb - FM 197-202; POH-4
   - ☐☐☐ Soft-field approach and landing - FM 203-207; POH-4
   - ☐☐☐ Short-field takeoff and maximum performance climb - FM 208-214; POH-4
   - ☐☐☐ Short-field approach and landing - FM 215-221; POH-4
   - ☐☐☐ Power-off 180° accuracy approach and landing - FM 222-230
   - ☐☐☐ Steep turns - FM 238-242
   - ☐☐☐ Steep spirals - FM 243-248
   - ☐☐☐ Chandelles - FM 249-253
   - ☐☐☐ Lazy eights - FM 254-259
   - ☐☐☐ Eights-on-pylons - FM 262-267
   - ☐☐☐ Additional items at CFI's discretion ______________________________

     ______________________________
4. Postflight critique and preview of next lesson

### Completion Standards

The lesson will have been successfully completed when the pilot completes each maneuver assigned for this solo flight.

**Instructor's comments:** ______________________________

______________________________

**Lesson assignment:** ______________________________

______________________________

**Notes:** ______________________________

______________________________

______________________________

______________________________

______________________________

______________________________

______________________________

______________________________

______________________________

______________________________

______________________________

______________________________

______________________________

______________________________

## FLIGHT LESSON 51: COMPLEX AIRPLANE--MANEUVERS REVIEW

### Objective

To review procedures and maneuvers covered previously.

### Text References

*Commercial Pilot Flight Maneuvers and Practical Test Prep* (FM)
Pilot's Operating Handbook (POH)

### Content

1. Flight Lesson 50 complete? Yes ___ Copy of lesson placed in pilot's folder? Yes ___
2. Preflight briefing
3. Review items
   - ☐☐☐ Operation of systems - FM 106-110; POH-1, 7, 8, 9
   - ☐☐☐ Supplemental oxygen - FM 326-327; POH-7, 8
   - ☐☐☐ Pressurization - FM 328-334; POH-7
   - ☐☐☐ Aeromedical factors - FM 111-116
   - ☐☐☐ Aeronautical decision making (ADM) - PH 343-349
   - ☐☐☐ Use of checklists - FM 11-12, 121; POH-4
   - ☐☐☐ Normal and crosswind takeoff and climb - FM 170-178; POH-4
   - ☐☐☐ Normal and crosswind approach and landing - FM 179-196; POH-4
   - ☐☐☐ Soft-field takeoff and climb - FM 197-202; POH-4
   - ☐☐☐ Soft-field approach and landing - FM 203-207; POH-4
   - ☐☐☐ Short-field takeoff and maximum performance climb - FM 208-214; POH-4
   - ☐☐☐ Short-field approach and landing - FM 215-221; POH-4
   - ☐☐☐ Power-off 180° accuracy approach and landing - FM 222-230
   - ☐☐☐ Traffic patterns - FM 153-159
     - ☐☐☐ Wake turbulence avoidance - FM 154-155
     - ☐☐☐ Runway incursion and collision avoidance - FM 153-154
     - ☐☐☐ Wind shear avoidance - FM 155
   - ☐☐☐ Go-around/rejected landing - FM 231-235; POH-4
   - ☐☐☐ Steep turns - FM 238-242
   - ☐☐☐ Steep spirals - FM 243-248
   - ☐☐☐ Chandelles - FM 249-253
   - ☐☐☐ Lazy eights - FM 254-259
   - ☐☐☐ Eights-on-pylons - FM 262-267
   - ☐☐☐ Emergency descent - FM 311-313; POH-3
   - ☐☐☐ Emergency approach and landing (simulated) - FM 314-320; POH-3
   - ☐☐☐ Systems and equipment malfunctions - FM 321-322; POH-3, 9
   - ☐☐☐ Additional items at CFI's discretion ________________________________

     ________________________________
4. Postflight critique and preview of next lesson

### Completion Standards

The lesson will have been successfully completed when the pilot demonstrates an understanding of the assigned procedures and maneuvers. The pilot will complete each task to the minimum standards listed in the current FAA Commercial Pilot Practical Test Standards. Those procedures and maneuvers not meeting the commercial standards will be thoroughly explained by the instructor and assigned for additional practice in the next lesson.

**Instructor's comments:** ________________________________

________________________________

**Lesson assignment:** ________________________________

________________________________

**Notes:** ________________________________

________________________________

## FLIGHT LESSON 52: SOLO PRACTICE

### Objective

To increase the pilot's proficiency in each of the assigned maneuvers. Additionally, the pilot will use this solo practice to correct any problem areas from Flight Lesson 51.

### Text References

*Commercial Pilot Flight Maneuvers and Practical Test Prep* (FM)
*Pilot Handbook* (PH)
Pilot's Operating Handbook (POH)

### Content

1. Flight Lesson 51 complete? Yes ____ Copy of lesson placed in pilot's folder? Yes ____
2. Preflight briefing
3. Review items
   - ☐☐☐ Soft-field takeoff and climb - FM 197-202; POH-4
   - ☐☐☐ Soft-field approach and landing - FM 203-207; POH-4
   - ☐☐☐ Short-field takeoff and maximum performance climb - FM 208-214; POH-4
   - ☐☐☐ Short-field approach and landing - FM 215-221; POH-4
   - ☐☐☐ Power-off 180° accuracy approach and landing - FM 222-230
   - ☐☐☐ Steep turns - FM 238-242
   - ☐☐☐ Steep spirals - FM 243-248
   - ☐☐☐ Chandelles - FM 249-253
   - ☐☐☐ Lazy eights - FM 254-259
   - ☐☐☐ Eights-on-pylons - FM 262-267
   - ☐☐☐ Power-off stalls - FM 289-294; PH 71-75
   - ☐☐☐ Power-on stalls - FM 295-300; PH 71-75
   - ☐☐☐ Accelerated stalls - FM 301-305; PH 71-75
   - ☐☐☐ Maneuvers as assigned by the instructor - CFI
   - ☐☐☐ Additional items at CFI's discretion ________________________________________

     ________________________________________
4. Postflight critique and preview of next lesson

### Completion Standards

The lesson will have been successfully completed when the pilot completes each maneuver assigned for this solo flight. Additionally, the pilot will attempt to correct the problem areas from Flight Lesson 51.

**Instructor's comments:** ________________________________________

________________________________________

**Lesson assignment:** ________________________________________

________________________________________

**Notes:** ________________________________________

________________________________________

________________________________________

________________________________________

________________________________________

________________________________________

________________________________________

________________________________________

________________________________________

## FLIGHT LESSON 53: MANEUVERS REVIEW

### Objective

To review procedures and maneuvers covered previously.

### Text References

*Commercial Pilot Flight Maneuvers and Practical Test Prep* (FM)
*Pilot Handbook* (PH)
Pilot's Operating Handbook (POH)

### Content

1. Flight Lesson 52 complete? Yes ___ Copy of lesson placed in pilot's folder? Yes ___
2. Preflight briefing
3. Review items
   - ☐☐☐ Radio communications and ATC light signals - FM 150-152
   - ☐☐☐ Airport, runway, and taxiway signs, markings, and lighting - FM 160-167
   - ☐☐☐ Runway incursion avoidance - FM 139-143
   - ☐☐☐ Soft-field takeoff and climb - FM 197-202; POH-4
   - ☐☐☐ Soft-field approach and landing - FM 203-207; POH-4
   - ☐☐☐ Short-field takeoff and maximum performance climb - FM 208-214; POH-4
   - ☐☐☐ Short-field approach and landing - FM 215-221; POH-4
   - ☐☐☐ Power-off 180° accuracy approach and landing - FM 222-230
   - ☐☐☐ Chandelles - FM 249-253
   - ☐☐☐ Lazy eights - FM 254-259
   - ☐☐☐ Eights-on-pylons - FM 262-267
   - ☐☐☐ Maneuvering during slow flight - FM 284-288
   - ☐☐☐ Power-off stalls - FM 289-294; PH 71-75
   - ☐☐☐ Power-on stalls - FM 295-300; PH 71-75
   - ☐☐☐ Accelerated stalls - FM 301-305; PH 71-75
   - ☐☐☐ Spin awareness - FM 306-308; POH-2
   - ☐☐☐ Recovery from unusual flight attitudes (IR) - CFI
   - ☐☐☐ Attitude instrument flying (IR) - CFI
   - ☐☐☐ Partial panel flying (IR) - CFI
   - ☐☐☐ Intercepting and tracking navigation systems (IR) - CFI
   - ☐☐☐ Additional items at CFI's discretion ______________________________

     ______________________________
4. Postflight critique and preview of next lesson

### Completion Standards

The lesson will have been successfully completed when the pilot demonstrates increased proficiency in the assigned maneuvers. The pilot will be able to complete each task to the minimum standards listed in the current FAA Commercial Pilot Practical Test Standards.

**Instructor's comments:** ______________________________

______________________________

**Lesson assignment:** ______________________________

______________________________

**Notes:** ______________________________

______________________________

______________________________

______________________________

______________________________

______________________________

______________________________

## FLIGHT LESSON 54: DUAL CROSS-COUNTRY

### Objective

To re-evaluate the pilot's ability to conduct a cross-country flight for the purpose of ensuring the pilot meets the minimum standards required by the current FAA Commercial Pilot Practical Test Standards. At a minimum, this flight will include a landing at an airport that is a straight-line distance of more than 50 NM from the original point of departure.

### Text References

*Commercial Pilot Flight Maneuvers and Practical Test Prep* (FM)
*Pilot Handbook* (PH)
Pilot's Operating Handbook (POH)

### Content

1. Flight Lesson 53 complete? Yes ____ Copy of lesson placed in pilot's folder? Yes ____
2. Preflight briefing
3. Review items

☐☐☐ Certificates and documents - FM 68-71
☐☐☐ Airworthiness requirements - FM 72-76
  ☐☐☐ Minimum equipment list (MEL) - FM 73-74; CFI
☐☐☐ Cross-country flight planning - FM 92-94
☐☐☐ Weather information - FM 77-91
☐☐☐ National Airspace System - FM 95-101
☐☐☐ Performance and limitations - FM 102-105; POH-2, 5, 6
☐☐☐ Operation of systems - FM 106-110; POH-1, 7, 8, 9
☐☐☐ Supplemental oxygen - FM 326-327; POH-7, 8
☐☐☐ Pressurization - FM 328-334; POH-7
☐☐☐ Preflight inspection - FM 118-122; POH-4
☐☐☐ Cockpit management - FM 123-126
☐☐☐ Engine starting - FM 127-131; POH-4
☐☐☐ Taxiing - FM 132-138
☐☐☐ Runway incursion avoidance - FM 139-143
☐☐☐ Before-takeoff check - FM 144-147; POH-4
☐☐☐ Radio communications and ATC light signals - FM 150-152
☐☐☐ Traffic patterns - FM 153-159
☐☐☐ Airport, runway, and taxiway signs, markings, and lighting - FM 160-167
☐☐☐ Soft-field takeoff and climb - FM 197-202; POH-4
☐☐☐ Soft-field approach and landing - FM 203-207; POH-4
☐☐☐ Short-field takeoff and maximum performance climb - FM 208-214; POH-4
☐☐☐ Short-field approach and landing - FM 215-221; POH-4
☐☐☐ Systems and equipment malfunctions - FM 321-322; POH-3, 9
☐☐☐ Attitude instrument flying (IR) - CFI
☐☐☐ Recovery from unusual attitudes (IR) - CFI
☐☐☐ Partial panel flying (IR) - CFI
☐☐☐ Intercepting and tracking navigational systems (IR) - CFI
☐☐☐ Pilotage and dead reckoning - FM 270-274
☐☐☐ Navigation systems and radar services - FM 275-276; PH 481-514
☐☐☐ Diversion - FM 277-279
☐☐☐ Lost procedures - FM 280-282
☐☐☐ Magnetic compass turns - FM 272-274; PH 96-101
☐☐☐ After landing, parking, and securing - FM 336-339; POH-4
☐☐☐ Additional items at CFI's discretion ____________
____________

4. Postflight critique and preview of next lesson

NOTE: This lesson may be completed under VFR or IFR at the discretion of the flight instructor and/or chief flight instructor, as appropriate.

### Completion Standards

The lesson will have been successfully completed when the pilot demonstrates the knowledge and skill to safely act as pilot in command during all phases of a cross-country flight. The pilot will be able to complete each task to the minimum standards listed in the current FAA Commercial Pilot Practical Test Standards.

**Instructor's comments:** ____________

____________

**Lesson assignment:** ____________

____________

**Notes:** ____________

____________

## FLIGHT LESSON 55: SOLO PRACTICE

### Objective

To increase the pilot's proficiency in preparation for the commercial pilot practical test.

### Text References

*Commercial Pilot Flight Maneuvers and Practical Test Prep* (FM)
*Pilot Handbook* (PH)
Pilot's Operating Handbook (POH)

### Content

1. Flight Lesson 54 complete? Yes ____ Copy of lesson placed in pilot's folder? Yes ____
2. Preflight briefing
3. Review items
   - ☐☐☐ Soft-field takeoff and climb - FM 197-202; POH-4
   - ☐☐☐ Soft-field approach and landing - FM 203-207; POH-4
   - ☐☐☐ Short-field takeoff and maximum performance climb - FM 208-214; POH-4
   - ☐☐☐ Short-field approach and landing - FM 215-221; POH-4
   - ☐☐☐ Power-off 180° accuracy approach and landing - FM 222-230
   - ☐☐☐ Steep turns - FM 238-242
   - ☐☐☐ Steep spirals - FM 243-248
   - ☐☐☐ Chandelles - FM 249-253
   - ☐☐☐ Lazy eights - FM 254-259
   - ☐☐☐ Eights-on-pylons - FM 262-267
   - ☐☐☐ Maneuvering during slow flight - FM 284-288
   - ☐☐☐ Power-off stalls - FM 289-294; PH 71-75
   - ☐☐☐ Power-on stalls - FM 295-300; PH 71-75
   - ☐☐☐ Accelerated stalls - FM 301-305; PH 71-75
   - ☐☐☐ Additional items at CFI's discretion ________________________________

     ________________________________
4. Postflight critique and preview of next lesson

### Completion Standards

The lesson will have been successfully completed when the pilot completes each maneuver assigned for this solo flight.

**Instructor's comments:** ________________________________

________________________________

**Lesson assignment:** ________________________________

________________________________

**Notes:** ________________________________

________________________________

________________________________

________________________________

________________________________

________________________________

________________________________

________________________________

________________________________

________________________________

________________________________

## FLIGHT LESSON 56: MANEUVERS REVIEW

### Objective

To review procedures and maneuvers covered previously.

### Text References

*Commercial Pilot Flight Maneuvers and Practical Test Prep* (FM)
*Pilot Handbook* (PH)
Pilot's Operating Handbook (POH)

### Content

1. Flight Lesson 55 complete? Yes ____ Copy of lesson placed in pilot's folder? Yes ____
2. Preflight briefing
3. Review items
   - ☐☐☐ Operation of systems - FM 106-110; POH-1, 7, 8, 9
   - ☐☐☐ Aeromedical factors - FM 111-116
   - ☐☐☐ Physiological aspects of night flying - CFI
   - ☐☐☐ Lighting and equipment for night flight - CFI
   - ☐☐☐ Normal and crosswind takeoff and climb - FM 170-178; POH-4
   - ☐☐☐ Normal and crosswind approach and landing - FM 179-196; POH-4
   - ☐☐☐ Soft-field takeoff and climb - FM 197-202; POH-4
   - ☐☐☐ Soft-field approach and landing - FM 203-207; POH-4
   - ☐☐☐ Short-field takeoff and maximum performance climb - FM 208-214; POH-4
   - ☐☐☐ Short-field approach and landing - FM 215-221; POH-4
   - ☐☐☐ Power-off 180° accuracy approach and landing - FM 222-230
   - ☐☐☐ Go-around/rejected landing - FM 231-235; POH-4
   - ☐☐☐ Lazy eights - FM 254-259
   - ☐☐☐ Maneuvering during slow flight - FM 284-288
   - ☐☐☐ Power-off stalls - FM 289-294; PH 71-75
   - ☐☐☐ Power-on stalls - FM 295-300; PH 71-75
   - ☐☐☐ Accelerated stalls - FM 301-305; PH 71-75
   - ☐☐☐ Spin awareness - FM 306-308; POH-2
   - ☐☐☐ Emergency descent - FM 311-313; POH-3
   - ☐☐☐ Emergency approach and landing (simulated) - FM 314-320; POH-3
   - ☐☐☐ Additional items at CFI's discretion ______________________________

     ______________________________
4. Postflight critique and preview of next lesson

### Completion Standards

The lesson will have been successfully completed when the pilot demonstrates proficiency in the maneuvers performed. The pilot will complete each task to the standards specified in the current FAA Commercial Pilot Practical Test Standards.

**Instructor's comments:** ______________________________

______________________________

**Lesson assignment:** ______________________________

______________________________

**Notes:** ______________________________

______________________________

______________________________

______________________________

______________________________

______________________________

______________________________

## FLIGHT LESSON 57: SOLO PRACTICE

### Objective

To increase the pilot's proficiency and confidence in preparation for the Stage Three Check flight.

### Text References

*Commercial Pilot Flight Maneuvers and Practical Test Prep* (FM)
Pilot's Operating Handbook (POH)

### Content

1. Flight Lesson 56 complete? Yes ____ Copy of lesson placed in pilot's folder? Yes ____
2. Preflight briefing
3. Review items
   - ☐☐☐ Soft-field takeoff and climb - FM 197-202; POH-4
   - ☐☐☐ Soft-field approach and landing - FM 203-207; POH-4
   - ☐☐☐ Short-field takeoff and maximum performance climb - FM 208-214; POH-4
   - ☐☐☐ Short-field approach and landing - FM 215-221; POH-4
   - ☐☐☐ Power-off 180° accuracy approach and landing - FM 222-230
   - ☐☐☐ Steep turns - FM 238-242
   - ☐☐☐ Steep spirals - FM 243-248
   - ☐☐☐ Chandelles - FM 249-253
   - ☐☐☐ Lazy eights - FM 254-259
   - ☐☐☐ Eights-on-pylons - FM 262-267
   - ☐☐☐ Maneuvers as assigned by the instructor - CFI
   - ☐☐☐ Additional items at CFI's discretion ________________________________

     ________________________________________________
4. Postflight critique and preview of next lesson

### Completion Standards

The lesson will have been successfully completed when the pilot completes each maneuver assigned for this solo flight.

**Instructor's comments:** ________________________________________________

________________________________________________

**Lesson assignment:** ________________________________________________

________________________________________________

**Notes:** ________________________________________________

________________________________________________

________________________________________________

________________________________________________

________________________________________________

________________________________________________

________________________________________________

________________________________________________

________________________________________________

________________________________________________

________________________________________________

________________________________________________

________________________________________________

## FLIGHT LESSON 58: COMPLEX AIRPLANE--MANEUVERS REVIEW

### Objective

To review procedures and maneuvers covered previously.

### Text References

*Commercial Pilot Flight Maneuvers and Practical Test Prep* (FM)
Pilot's Operating Handbook (POH)

### Content

1. Flight Lesson 57 complete? Yes ____ Copy of lesson placed in pilot's folder? Yes ____
2. Preflight briefing
3. Review items
   - ☐☐☐ Operation of systems - FM 106-110; POH-1, 7, 8, 9
   - ☐☐☐ Supplemental oxygen - FM 326-327; POH-7, 8
   - ☐☐☐ Pressurization - FM 328-334; POH-7
   - ☐☐☐ Runway incursion avoidance - FM 139-143
   - ☐☐☐ Soft-field takeoff and climb - FM 197-202; POH-4
   - ☐☐☐ Soft-field approach and landing - FM 203-207; POH-4
   - ☐☐☐ Short-field takeoff and maximum performance climb - FM 208-214; POH-4
   - ☐☐☐ Short-field approach and landing - FM 215-221; POH-4
   - ☐☐☐ Power-off 180° accuracy approach and landing - FM 222-230
   - ☐☐☐ Traffic patterns - FM 153-159
   - ☐☐☐ Go-around/rejected landing - FM 231-235; POH-4
   - ☐☐☐ Steep turns - FM 238-242
   - ☐☐☐ Steep spirals - FM 243-248
   - ☐☐☐ Chandelles - FM 249-253
   - ☐☐☐ Lazy eights - FM 254-259
   - ☐☐☐ Eights-on-pylons - FM 262-267
   - ☐☐☐ Emergency descent - FM 311-313; POH-3
   - ☐☐☐ Emergency approach and landing (simulated) - FM 314-320; POH-3
   - ☐☐☐ Systems and equipment malfunctions - FM 321-322; POH-3, 9
   - ☐☐☐ Emergency equipment and survival gear - FM 323-324; POH-7, 9
   - ☐☐☐ Additional items at CFI's discretion ____________________

     ____________________
4. Postflight critique and preview of next lesson

### Completion Standards

The lesson will have been successfully completed when the pilot demonstrates proficiency at the commercial pilot skill level in all the assigned procedures and maneuvers. The pilot will complete each task to the standards specified in the current FAA Commercial Pilot Practical Test Standards.

**Instructor's comments:** ____________________

____________________

**Lesson assignment:** ____________________

____________________

**Notes:** ____________________

____________________

____________________

____________________

____________________

____________________

____________________

## FLIGHT LESSON 59: STAGE THREE CHECK--COMPLEX AIRPLANE

### Objective

During this stage check, an authorized instructor will determine if the pilot is proficient in the commercial pilot flight maneuvers.

### Text References

*Commercial Pilot Flight Maneuvers and Practical Test Prep* (FM)
*Pilot Handbook* (PH)
Pilot's Operating Handbook (POH)

### Content

1. Flight Lesson 58 complete? Yes ____ Copy of lesson placed in pilot's folder? Yes ____
2. Preflight briefing
3. Stage check tasks

☐☐☐ Performance and limitations - FM 102-105; POH-2, 5, 6
☐☐☐ Operation of systems - FM 106-110; POH-1, 7, 8, 9
☐☐☐ Preflight inspection - FM 118-122; POH-4
☐☐☐ Cockpit management - FM 123-126
☐☐☐ Engine starting - FM 127-131; POH-4
☐☐☐ Taxiing - FM 132-138
☐☐☐ Runway incursion avoidance - FM 139-143
☐☐☐ Before-takeoff check - FM 144-147; POH-4
☐☐☐ Radio communications and ATC light signals - FM 150-152
☐☐☐ Traffic patterns - FM 153-159
☐☐☐ Airport, runway, and taxiway signs, markings, and lighting - FM 160-167
☐☐☐ Normal and crosswind takeoff and climb - FM 170-178; POH-4
☐☐☐ Normal and crosswind approach and landing - FM 179-196; POH-4
☐☐☐ Soft-field takeoff and climb - FM 197-202; POH-4
☐☐☐ Soft-field approach and landing - FM 203-207; POH-4
☐☐☐ Short-field takeoff and maximum performance climb - FM 208-214; POH-4
☐☐☐ Short-field approach and landing - FM 215-221; POH-4
☐☐☐ Power-off 180° accuracy approach and landing - FM 222-230
☐☐☐ Go-around/rejected landing - FM 231-235; POH-4
☐☐☐ Steep turns - FM 238-242
☐☐☐ Steep spirals - FM 243-248
☐☐☐ Chandelles - FM 249-253
☐☐☐ Lazy eights - FM 254-259
☐☐☐ Eights-on-pylons - FM 262-267
☐☐☐ Maneuvering during slow flight - FM 284-288
☐☐☐ Power-off stalls - FM 289-294; PH 71-75
☐☐☐ Power-on stalls - FM 295-300; PH 71-75
☐☐☐ Accelerated stalls - FM 301-305; PH 71-75
☐☐☐ Spin awareness - FM 306-308; POH-2
☐☐☐ Straight-and-level (IR) - CFI
☐☐☐ Constant airspeed climbs and descents (IR) - CFI
☐☐☐ Turns to a heading (IR) - CFI
☐☐☐ Recovery from unusual attitudes (IR) - CFI
☐☐☐ Emergency descent - FM 311-313; POH-3
☐☐☐ Emergency approach and landing (simulated) - FM 314-320; POH-3
☐☐☐ Systems and equipment malfunctions - FM 321-322; POH-3, 9
☐☐☐ Emergency equipment and survival gear - FM 323-324; POH-7, 9
☐☐☐ After landing, parking, and securing - FM 336-339; POH-4
☐☐☐ Additional items at CFI's discretion ____________

______________________________

______________________________

4. Postflight critique

### Completion Standards

This lesson and Stage Three will have been successfully completed when the pilot demonstrates the required level of proficiency in all stage check tasks based on the current FAA Commercial Pilot Test Standards. If additional instruction is necessary, the check instructor will assign the additional training. If the flight is satisfactory, the check instructor will complete the pilot's training records and recommend the pilot for the End-of-Course check.

**Instructor's comments:** ______________________________

______________________________

**Lesson assignment:** ______________________________

______________________________

**Notes:** ______________________________

______________________________

______________________________

## FLIGHT LESSON 60: END-OF-COURSE CHECK--COMPLEX AIRPLANE

### Objective

The pilot will be able to demonstrate the required proficiency of a commercial pilot by utilizing the current FAA Commercial Pilot Practical Test Standards. Note: If a noncomplex and a complex airplane are used, a complex airplane must be used for all takeoffs, landings, and emergency procedures.

### Text References

*Commercial Pilot Flight Maneuvers and Practical Test Prep* (FM)
*Pilot Handbook* (PH)
Pilot's Operating Handbook (POH)

### Content

1. Flight Lesson 59 complete? Yes ____ Copy of lesson placed in pilot's folder? Yes ____
2. Preflight briefing
3. Stage check tasks

☐☐☐ Certificates and documents - FM 68-71
☐☐☐ Airworthiness requirements - FM 72-76
☐☐☐ Weather information - FM 77-91
☐☐☐ Cross-country flight planning - FM 92-94
☐☐☐ National Airspace System - FM 95-101
☐☐☐ Performance and limitations - FM 102-105; POH-2, 5, 6
☐☐☐ Operation of systems - FM 106-110; POH-1, 7, 8, 9
☐☐☐ Aeromedical factors - FM 111-116
☐☐☐ Physiological aspects of night flying - CFI
☐☐☐ Lighting and equipment for night flying - CFI
☐☐☐ Preflight inspection - FM 118-122; POH-4
☐☐☐ Cockpit management - FM 123-126
☐☐☐ Engine starting - FM 127-131; POH-4
☐☐☐ Taxiing - FM 132-138
☐☐☐ Runway incursion avoidance - FM 139-143
☐☐☐ Before-takeoff check - FM 144-147; POH-4
☐☐☐ Radio communications and ATC light signals - FM 150-152
☐☐☐ Traffic patterns - FM 153-159
☐☐☐ Airport, runway, and taxiway signs, markings, and lighting - FM 160-167
☐☐☐ Normal and crosswind takeoff and climb - FM 170-178; POH-4
☐☐☐ Normal and crosswind approach and landing - FM 179-196; POH-4
☐☐☐ Soft-field takeoff and climb - FM 197-202; POH-4
☐☐☐ Soft-field approach and landing - FM 203-207; POH-4
☐☐☐ Short-field takeoff and maximum performance climb - FM 208-214; POH-4
☐☐☐ Short-field approach and landing - FM 215-221; POH-4
☐☐☐ Power-off 180° accuracy approach and landing - FM 222-230
☐☐☐ Go-around/rejected landing - FM 231-235; POH-4
☐☐☐ Steep turns - FM 238-242
☐☐☐ Steep spirals - FM 243-248
☐☐☐ Chandelles - FM 249-253
☐☐☐ Lazy eights - FM 254-259
☐☐☐ Eights-on-pylons - FM 262-267
☐☐☐ Pilotage and dead reckoning - FM 270-274
☐☐☐ Navigation systems and radar services - FM 275-276; PH 481-514
☐☐☐ Diversion - FM 277-279
☐☐☐ Lost procedures - FM 280-282
☐☐☐ Magnetic compass turns - FM 272-274; PH 96-101
☐☐☐ Maneuvering during slow flight - FM 284-288
☐☐☐ Power-off stalls - FM 289-294; PH 71-75
☐☐☐ Power-on stalls - FM 295-300; PH 71-75
☐☐☐ Accelerated stalls - FM 301-305; PH 71-75
☐☐☐ Spin awareness - FM 306-308; POH-2
☐☐☐ Emergency descent - FM 311-313; POH-3
☐☐☐ Emergency approach and landing (simulated) - FM 314-320; POH-3
☐☐☐ Systems and equipment malfunctions - FM 321-322; POH-3, 9
☐☐☐ Emergency equipment and survival gear - FM 323-324; POH-7, 9
☐☐☐ Supplemental oxygen - FM 326-327; POH-7, 8
☐☐☐ Pressurization - FM 328-334; POH-7
☐☐☐ After landing, parking, and securing - FM 336-339; POH-4
☐☐☐ Additional items at CFI's discretion ____________
____________________________________

4. Flight Lesson 60 complete? Yes ____ Copy of graduation certificate placed in pilot's folder? Yes ____
5. Postflight critique

### Completion Standards

This lesson and the course of training will have been successfully completed when the pilot demonstrates the required level of proficiency in all tasks of the current FAA Commercial Pilot Test Standards. If additional instruction is necessary, the chief flight instructor will assign the additional training. If the flight is satisfactory, the chief flight instructor will complete the pilot's training records and issue a graduation certificate.

**Instructor's comments:** ____________________________________

**Notes:** ____________________________________

# PART II: COMMERCIAL PILOT TRAINING SYLLABUS AIRPLANE MULTI-ENGINE LAND ADD-ON RATING



## PART 141 STUDENT INFORMATION

### Enrollment Prerequisites

You must hold at least a commercial pilot certificate, with an airplane category and single-engine land rating, and an instrument rating, with an airplane category, prior to enrolling in the flight portion of the commercial pilot multi-engine add-on rating certification course. If you do not hold an instrument rating at the time of enrollment, you may continue the course if you are concurrently enrolled in an approved instrument rating course.

### Graduation Requirements

You must complete the training specified in this syllabus, with a minimum of 15 hours of ground training in the specified aeronautical knowledge areas and a minimum of 15 hours of flight training. These requirements are reflected in the Gleim ground and flight training syllabus.

### Stage Checks

You must score a minimum of 80% on the comprehensive knowledge test at the conclusion of the ground training.

You must satisfactorily complete a comprehensive end-of-course flight check when all flight training is complete.

### Credit for Previous Training

You may be given credit toward this commercial pilot multi-engine add-on rating course for previous pilot experience and knowledge [14 CFR 141.77(c)]:

1. If the credit is based on a Part 141 training course, the credit may be 50% of the requirements for this course.
2. If the credit is based on a Part 61 course, the credit cannot exceed 25% of the requirements for this course.

The receiving school will determine the amount of course credit to be given based on a proficiency test, a knowledge test, or both.

## STRUCTURE OF THE GLEIM COMMERCIAL PILOT SYLLABUS (MULTI-ENGINE ADD-ON)

This syllabus consists of a ground training syllabus and a flight training syllabus. The ground and flight training may be done together as an integrated course of instruction, or each may be done separately. If done separately, the ground syllabus may be conducted as a home-study course or as a formal ground school.

This syllabus was constructed using the building-block progression of learning, in which the student is required to perform each task correctly before a more complex task is introduced. This method will promote the formation of correct habit patterns from the beginning.

### Ground Training Syllabus

The ground training syllabus contains nine lessons. The ground training syllabus meets the training requirements of Appendix D to Part 141 and Sec. 61.125. The ground training can be conducted concurrently with the flight training, with the ground lessons completed in order as outlined in the lesson matrix. Ground training may also be conducted as part of a formal ground school or as a home-study program.

It is recommended that the lessons be completed in sequence, but the syllabus is flexible enough to meet the needs of an individual student or of a particular training environment. When departing from the sequence, the instructor is responsible for considering the blocks of learning affected and, if used by a Part 141 pilot school, whether it would affect FAA approval.

Each ground lesson involves studying the appropriate content from the Gleim **Multi-Engine Add-On Rating Course**, delivered online. At the end of each lesson, you will take a quiz or be quizzed by your instructor. You will review any incorrect responses with your instructor to ensure understanding.

An end-of-course knowledge test (found in Appendix B of this book) is administered after Ground Lesson 9 is completed. Upon successful completion of the end-of-course knowledge test, you will have successfully completed this ground training syllabus.

If this ground training is used as home study, we recommend that you complete the syllabus as quickly as possible so you will have more time to prepare for your flight lessons and be able to expedite your flight training.

NOTE: We recommend that you complete all of the ground lessons before beginning your multi-engine flight training. You may elect to follow the lesson sequence shown on page 104, but we recommend that you use these ground lessons as an opportunity to review the lesson content with your flight instructor. By studying the material on your own, you will develop a better sense of the content, be better informed during your ground training, and ask better questions during your sessions with your instructor.

## Flight Training Syllabus

The flight training syllabus contains 10 lessons. This course of training is designed to expose you to multi-engine airplane operations in an ordered, logical sequence, which will aid your ability to transition to multi-engine flying faster. It is recommended that each lesson be completed in sequential order.

**Stage check.** Stage checks are designed to ensure that the student has acquired the necessary knowledge and skill. The End-of-Course check (Lesson 10) is designed to ensure that you are proficient in all multi-engine knowledge and skill areas and are ready for the commercial pilot multi-engine add-on practical test. It is also the final evaluation for a Part 141 graduation certificate. The chief flight instructor (Part 141) is responsible for ensuring that each pilot accomplishes the required end-of-course stage test.

The End-of-Course check should also be used as a review by instructors training under Part 61 to ensure that the student has the appropriate knowledge and skills.

**Sequence of a flight lesson.** Each flight lesson will begin with a preflight briefing. During this time, the instructor should first answer any questions you may have from the previous lesson. Next, your instructor will brief you on the lesson content, followed by an evaluation of your preparation for the lesson.

During the flight portion of the lesson, your instructor should begin with those maneuvers listed as "review items" before introducing new maneuvers. The time required for each lesson will vary depending on the airport and the location of the training areas.

At the end of each lesson, your instructor will conduct a postflight critique and a preview of the next lesson. This time should be used to review what went well during the lesson and to identify and discuss any problem areas so that you can correct any deficiencies on subsequent flights.

The length of the preflight briefing and postflight critique will vary with each student and with his/her degree of preparedness for the lesson.

**Pilot preparation.** The key to minimizing frustration and cost is preparation. You should budget an average of 2 to 4 hours of home study prior to each flight lesson. Learning will be easier when you are fully prepared so that your instructor can maximize the time spent in flight training.

## MULTI-ENGINE ADD-ON RATING SYLLABUS LESSON SEQUENCE AND TIMES

The Ground Syllabus follows on pages 105 through 116, and the Flight Syllabus follows on pages 117 through 129.

The table below lists the sequence of the flight and ground lessons and the minimum time for each lesson. The times listed are for instructor/student guidance only and are not meant to be mandatory times. These times will ensure that the minimum time requirements for aeronautical knowledge and flight training are in compliance with Part 141, Appendix I, Additional Aircraft Category and/or Class Rating Course. Refer to our note on ground training on page 102 regarding accomplishing the ground portion of this training course.

The major difference between Part 141 and Part 61 training for the multi-engine add-on rating is that Part 61 requires no minimum amount of flight time or ground training to add the rating onto an existing commercial pilot certificate. This syllabus encourages a safety mindset and is more than adequate to meet the training needs of Part 61.

Each training flight must include a preflight briefing and a postflight critique of the pilot's performance by the instructor.

| | | Flight | | | | | | | | | Ground |
|---|---|---|---|---|---|---|---|---|---|---|---|
| LESSON | Page | Flight Training Dual | Flight Training Solo | Dual Cross-Country | Solo Cross-Country | Instrument | Night Dual | Night Solo | Complex Airplane Dual | Complex Airplane Solo | Aeronautical Knowledge Training |
| Ground 1 Multi Add-On | 107 | | | | | | | | | | 1.0 |
| Ground 2 Aerodynamics | 108 | | | | | | | | | | 2.0 |
| Flight 1 Intro to Multi-Engine Airplanes | 120 | 1.0 | | | | | | | 1.0 | | |
| Ground 3 Systems | 109 | | | | | | | | | | 2.0 |
| Ground 4 Safety Considerations | 110 | | | | | | | | | | 1.0 |
| Flight 2 Normal Maneuvers | 121 | 1.0 | | | | | | | 1.0 | | |
| Ground 5 Performance, Wt. & Bal. | 111 | | | | | | | | | | 1.5 |
| Ground 6 Normal Operations | 112 | | | | | | | | | | 2.0 |
| Flight 3 Performance Maneuvers | 122 | 1.5 | | | | | | | 1.5 | | |
| Flight 4 Instrument Flying Skills | 123 | 1.5 | | | | 1.0 | | | 1.5 | | |
| Ground 7 Abnormal Operations | 113 | | | | | | | | | | 2.0 |
| Flight 5 Abnormal Maneuvers | 124 | 1.5 | | | | 0.5 | | | 1.5 | | |
| Flight 6 Emergency Operations | 125 | 1.5 | | | | 0.5 | | | 1.5 | | |
| Flight 7 Day Cross-Country | 126 | 2.0 | | 2.0 | | 1.0 | | | 2.0 | | |
| Flight 8 Night Cross-Country | 127 | 2.0 | | 2.0 | | 1.0 | 2.0 | | 2.0 | | |
| Ground 8 Your Practical Test | 114 | | | | | | | | | | 1.5 |
| Ground 9 Oral Exam Guide | 115 | | | | | | | | | | 1.5 |
| Flight 9 Maneuvers Review | 128 | 1.5 | | | | 0.5 | | | 1.5 | | |
| Ground End-of-Course Knowledge Test | 116 | | | | | | | | | | 0.5 |
| Flight 10 End-Of-Course Check | 129 | 1.5 | | | | 0.5 | | | 1.5 | | |
| **TOTALS** | | **15.0** | | **4.0** | | **5.0** | **2.0** | | **15.0** | | **15.0** |

# PART II: COMMERCIAL PILOT GROUND TRAINING SYLLABUS AIRPLANE MULTI-ENGINE LAND ADD-ON RATING

## GROUND TRAINING COURSE OBJECTIVES

The pilot will obtain the necessary aeronautical knowledge and meet the prerequisites specified in Appendix I to 14 CFR Part 141 (and the appropriate items from Sec. 61.125) to successfully pass the end-of-course knowledge test and subsequently the commercial pilot multi-engine add-on practical test.

## GROUND TRAINING COURSE COMPLETION STANDARDS

The pilot will demonstrate through individual ground lesson quizzing, a cumulative end-of-course knowledge test, and school records that (s)he meets the prerequisites specified in Appendix I to 14 CFR Part 141 (and the appropriate items from Sec. 61.125) and has the aeronautical knowledge necessary to apply for the commercial pilot multi-engine add-on practical test.

| Lesson | Topic | Min. Time in Hours | Reading Assignment No. of Pgs. | No. of Questions |
|---|---|---|---|---|
| 1 | The Multi-Engine Add-On Rating | 1.0 | 17 | 20 |
| 2 | Multi-Engine Aerodynamics | 2.0 | 24 | 20 |
| 3 | Multi-Engine Airplane Systems | 2.0 | 23 | 20 |
| 4 | Multi-Engine Airplane Safety Considerations | 1.0 | 24 | 20 |
| 5 | Multi-Engine Airplane Performance and Weight and Balance | 1.5 | 18 | 20 |
| 6 | Multi-Engine Airplane Normal Operations | 2.0 | 36 | 20 |
| 7 | Multi-Engine Airplane Abnormal and Emergency Operations | 2.0 | 21 | 20 |
| 8 | Your FAA Practical Test | 1.5 | 24 | 20 |
| 9 | Oral Exam Guide | 1.5 | 40 | N/A |
| | End-of-Course Knowledge Test | 0.5 | | |

## GROUND LESSON 1: THE MULTI-ENGINE ADD-ON RATING

### Objective

To further develop the pilot's knowledge of the multi-engine rating, the requirements for obtaining the rating, and the regulations that impact operating with the rating.

### Text References

Multi-Engine Add-On Rating Course, Study Unit 1, "The Multi-Engine Add-On Rating" (17 pages)

Multi-Engine Add-On Rating Course
Study Unit 1 Contents



### Completion Standards

The lesson will have been successfully completed when the pilot answers the true/false study questions, reviews the knowledge transfer outline, and passes the multiple-choice test for Study Unit 1 in the Gleim Multi-Engine Add-On Rating Course with a minimum passing grade of 80%.

---

| | Dates Studied | | | | | Date Completed |
|---|---|---|---|---|---|---|
| Multi-Engine Add-On Rating Course | _____ | _____ | _____ | _____ | _____ | _____ |

Notes:

## GROUND LESSON 2: MULTI-ENGINE AERODYNAMICS

### Objective

To further develop the pilot's knowledge of multi-engine aerodynamics as well as the performance and controllability problems associated with single-engine operations.

### Text References

Multi-Engine Add-On Rating Course, Study Unit 2, "Multi-Engine Aerodynamics" (24 pages)

Multi-Engine Add-On Rating Course
Study Unit 2 Contents



### Completion Standards

The lesson will have been successfully completed when the pilot answers the true/false study questions, reviews the knowledge transfer outline, and passes the multiple-choice test for Study Unit 2 in the Gleim Multi-Engine Add-On Rating Course with a minimum passing grade of 80%.

| | Dates Studied | | | | | Date Completed |
|---|---|---|---|---|---|---|
| Multi-Engine Add-On Rating Course | _____ | _____ | _____ | _____ | _____ | _____ |

Notes:

## GROUND LESSON 3: MULTI-ENGINE AIRPLANE SYSTEMS

### Objective

To further develop the pilot's knowledge of multi-engine airplane systems, especially systems that are unique in multi-engine applications.

### Text References

Multi-Engine Add-On Rating Course, Study Unit 3, "Multi-Engine Airplane Systems" (23 pages)

Multi-Engine Add-On Rating Course
Study Unit 3 Contents



### Completion Standards

The lesson will have been successfully completed when the pilot answers the true/false study questions, reviews the knowledge transfer outline, and passes the multiple-choice test for Study Unit 3 in the Gleim Multi-Engine Add-On Rating Course with a minimum passing grade of 80%.

---

| | Dates Studied | | | | | Date Completed |
|---|---|---|---|---|---|---|
| Multi-Engine Add-On Rating Course | _____ | _____ | _____ | _____ | _____ | _____ |

Notes:

## GROUND LESSON 4: MULTI-ENGINE AIRPLANE SAFETY CONSIDERATIONS

### Objective

To further develop the pilot's knowledge of basic aviation safety concepts, particularly those that have special impact on multi-engine operations.

### Text References

Multi-Engine Add-On Rating Course, Study Unit 4, "Multi-Engine Airplane Safety Considerations" (24 pages)

Multi-Engine Add-On Rating Course
Study Unit 4 Contents

*4.1 General*
*4.2 Air Traffic Control Emergency Services*
*4.3 Emergency ATC Communications*
*4.4 Radio Failure Procedures*
*4.5 Post-September 11, 2001, Considerations*
*4.6 Risk Management and Aeronautical Decision Making*
*4.7 Land and Hold Short Operations (LAHSO)*
*4.8 Avoiding Wake Turbulence*
*4.9 Avoiding Runway Incursions*
*4.10 Collision Avoidance*
*4.11 Airport Signs, Markings, and Lighting*
*4.12 Aviation Safety Reporting System (ASRS)*

### Completion Standards

The lesson will have been successfully completed when the pilot answers the true/false study questions, reviews the knowledge transfer outline, and passes the multiple-choice test for Study Unit 4 in the Gleim Multi-Engine Add-On Rating Course with a minimum passing grade of 80%.

---

| | Dates Studied | | | | | Date Completed |
|---|---|---|---|---|---|---|
| Multi-Engine Add-On Rating Course | ____ | ____ | ____ | ____ | ____ | ____ |

Notes:

## GROUND LESSON 5: MULTI-ENGINE AIRPLANE PERFORMANCE AND WEIGHT AND BALANCE

### Objective

To further develop the pilot's knowledge of multi-engine performance and weight-and-balance definitions, principles, and calculations.

### Text References

Multi-Engine Add-On Rating Course, Study Unit 5, "Multi-Engine Airplane Performance and Weight and Balance" (18 pages)

Multi-Engine Add-On Rating Course
Study Unit 5 Contents



### Completion Standards

The lesson will have been successfully completed when the pilot answers the true/false study questions, reviews the knowledge transfer outline, and passes the multiple-choice test for Study Unit 5 in the Gleim Multi-Engine Add-On Rating Course with a minimum passing grade of 80%.

---

| | Dates Studied | | | | | Date Completed |
|---|---|---|---|---|---|---|
| Multi-Engine Add-On Rating Course | ____ | ____ | ____ | ____ | ____ | ____ |

Notes:

## GROUND LESSON 6: MULTI-ENGINE AIRPLANE NORMAL OPERATIONS

### Objective

To further develop the pilot's knowledge of the multi-engine Practical Test Standards tasks associated with normal operations.

### Text References

Multi-Engine Add-On Rating Course, Study Unit 6, "Multi-Engine Airplane Normal Operations" (36 pages)

Multi-Engine Add-On Rating Course
Study Unit 6 Contents

*6.1 General*
*6.2 Preflight Inspection (Task II.A.)*
*6.3 Cockpit Management (Task II.B.)*
*6.4 Engine Starting (Task II.C.)*
*6.5 Taxiing (Task II.D.)*
*6.6 Runway Incursion Avoidance (Task II.F.)*
*6.7 Before Takeoff Check (Task II.G.)*
*6.8 Normal and Crosswind Takeoff and Climb (Task IV.A.)*
*6.9 Normal and Crosswind Approach and Landing (Task IV.B.)*
*6.10 Short-Field Takeoff and Maximum Performance Climb (Task IV.C.)*
*6.11 Short-Field Approach and Landing (Task IV.D.)*
*6.12 Steep Turns (Task V.A.)*
*6.13 Maneuvering During Slow Flight (Task VII.A.)*
*6.14 Power-Off Stalls (Task VII.B.)*
*6.15 Power-On Stalls (Task VII.C.)*
*6.16 Accelerated Stalls (Task VII.D.)*

### Completion Standards

The lesson will have been successfully completed when the pilot answers the true/false study questions, reviews the knowledge transfer outline, and passes the multiple-choice test for Study Unit 6 in the Gleim Multi-Engine Add-On Rating Course with a minimum passing grade of 80%.

---

| | Dates Studied | | | | | Date Completed |
|---|---|---|---|---|---|---|
| Multi-Engine Add-On Rating Course | ____ | ____ | ____ | ____ | ____ | ____ |

Notes:

## GROUND LESSON 7: MULTI-ENGINE AIRPLANE ABNORMAL AND EMERGENCY OPERATIONS

### Objective

To further develop the pilot's knowledge of the multi-engine Practical Test Standards tasks associated with abnormal operations.

### Text References

Multi-Engine Add-On Rating Course, Study Unit 7, "Multi-Engine Airplane Abnormal and Emergency Operations" (21 pages)

Multi-Engine Add-On Rating Course
Study Unit 7 Contents

- *7.1 General*
- *7.2 Emergency Descent (Task VIII.A.)*
- *7.3 Engine Failure During Takeoff Before $V_{MC}$ (Task VIII.B.)*
- *7.4 Engine Failure After Lift-Off (Task VIII.C.)*
- *7.5 Approach and Landing with an Inoperative Engine (Task VIII.D.)*
- *7.6 Systems and Equipment Malfunctions (Task VIII.E.)*
- *7.7 Maneuvering with One Engine Inoperative (Task X.A.)*
- *7.8 $V_{MC}$ Demonstration (Task X.B.)*
- *7.9 Engine Failure During Flight (By Reference to Instruments) (Task X.C.)*
- *7.10 Instrument Approach – One Engine Inoperative (By Reference to Instruments) (Task X.D.)*

### Completion Standards

The lesson will have been successfully completed when the pilot answers the true/false study questions, reviews the knowledge transfer outline, and passes the multiple-choice test for Study Unit 7 in the Gleim Multi-Engine Add-On Rating Course with a minimum passing grade of 80%.

---

| | Dates Studied | | | | | Date Completed |
|---|---|---|---|---|---|---|
| Multi-Engine Add-On Rating Course | _____ | _____ | _____ | _____ | _____ | _____ |

Notes:

## GROUND LESSON 8: YOUR FAA PRACTICAL TEST

### Objective

To further develop the pilot's knowledge of the process for completing the FAA multi-engine add-on practical test, as well as to review the knowledge tasks in the Practical Test Standards likely to be covered during the oral portion of the practical test.

### Text References

Multi-Engine Add-On Rating Course, Study Unit 8, "Your FAA Practical Test," Subunits 1-8, 10 (24 pages)

Multi-Engine Add-On Rating Course
Study Unit 8 Contents



### Completion Standards

The lesson will have been successfully completed when when the pilot answers the true/false study questions, reviews the knowledge transfer outline, and passes the multiple-choice test for Study Unit 8 in the Gleim Multi-Engine Add-On Rating Course with a minimum passing grade of 80%.

---

| | Dates Studied | | | | | Date Completed |
|---|---|---|---|---|---|---|
| Multi-Engine Add-On Rating Course | _____ | _____ | _____ | _____ | _____ | _____ |

Notes:

## GROUND LESSON 9: ORAL EXAM GUIDE

### Objective

To further develop the pilot's knowledge of the topics usually covered during the oral portion of the multi-engine airplane practical test and to prepare the pilot for the oral exam.

### Text References

Multi-Engine Add-On Rating Course, Study Unit 8, "Your FAA Practical Test," Subunit 9, Oral Exam Guide (40 pages)

| Multi-Engine Add-On Rating Course<br>Study Unit 8 Contents |
|---|
| *8.9 Oral Exam Guide* |

### Completion Standards

The lesson will have been successfully completed when the pilot reviews the Oral Exam Guide, provided as a link in Study Unit 8, Subunit 9, of the Gleim Multi-Engine Add-On Rating Course, and displays to his/her instructor a satisfactory level of understanding through oral quizzing according to the current Practical Test Standards.

| | Dates Studied | | | | | Date Completed |
|---|---|---|---|---|---|---|
| Multi-Engine Add-On Rating Course | _____ | _____ | _____ | _____ | _____ | _____ |

Notes:

## END-OF-COURSE KNOWLEDGE TEST

### Objective

To evaluate the pilot's knowledge on the aeronautical knowledge subject areas required by Appendix I to 14 CFR Part 141 and to determine if the pilot is prepared for the oral exam portion of the commercial pilot multi-engine add-on practical test.

### Content

Questions similar to those that may be provided during the commercial pilot multi-engine add-on practical test.

### Completion Standards

The lesson and the ground training syllabus will have been successfully completed when the pilot has completed this knowledge test with a minimum passing grade of 80%.

# PART II: COMMERCIAL PILOT FLIGHT TRAINING SYLLABUS AIRPLANE MULTI-ENGINE LAND ADD-ON RATING



## FLIGHT TRAINING COURSE OBJECTIVES

The pilot will obtain the aeronautical knowledge and experience and demonstrate the flight proficiency necessary to meet the requirements for a commercial pilot certificate with a multi-engine land class add-on rating.

## FLIGHT TRAINING COURSE COMPLETION STANDARDS

The pilot will demonstrate through the End-of-Course Check and school records that (s)he has the necessary flight proficiency and aeronautical experience to obtain a commercial pilot certificate with a multi-engine land class add-on rating.

| Lesson | Topic | Number of Pages in "New Item" Reading Assignment |
|---|---|---|
| 1 | Introduction to Multi-Engine Airplanes | 148 |
| 2 | Normal Multi-Engine Flight Maneuvers | 38 |
| 3 | Multi-Engine Performance Maneuvers | 37 |
| 4 | Instrument Flying Skills | 50 |
| 5 | Abnormal Multi-Engine Flight Maneuvers | 13 |
| 6 | Emergency Operations | 9 |
| 7 | Multi-Engine Day Cross-Country | 53 |
| 8 | Multi-Engine Night Cross-Country | * |
| 9 | Maneuvers Review | * |
| 10 | End-of-Course Check | * |
| | | 348 |

*Reading assignment consists of review only, no new materials.

The following is a brief description of the parts of each flight lesson in this syllabus:

**Objective:** We open each lesson with an objective, usually a sentence or two, to help you gain perspective and understand the goal for that particular lesson.

**Text References:** This section tells you which reference materials you will need to study or refer to while mastering the tasks within the lesson. Abbreviations are given to facilitate the cross-referencing process.

**Content:** Each lesson contains a list of the tasks required to be completed before moving to the next lesson. A task may be listed as a "review item" (a task that was covered in a previous lesson) or as a "new item" (a task that is introduced to you for the first time in this course). Each task is preceded by three blank "checkoff" boxes, which may be used by your instructor to keep track of your progress and to indicate that each task was completed.

There are three boxes because it may take more than one flight to complete the lesson. Your instructor may mark the box(es) next to each task in one of the following methods (or any other method desired):

| | | |
|---|---|---|
| ✓ - task completed to lesson completion standards | D - demonstrated by instructor *<br>A - accomplished by you<br>S - safe/satisfactory<br>P - meets PTS standards | 1 - above lesson standard<br>2 - meets lesson standard<br>3 - below lesson standard |

Most tasks are followed by book and page references that tell you where to find the information you need to study to accomplish the task successfully.

The last task in each flight lesson is labeled "Additional items at CFI's discretion" and is followed by several blank lines. This area can be used to record any extra items that your instructor feels are appropriate to the lesson, taking into account such variables as weather, local operational considerations, and your progress as a student.

NOTE: Instructors are reminded not to limit themselves to the blank lines provided -- use as much of the page as you need.

**Completion Standards:** Based on these standards, your instructor determines how well you have met the objective of the lesson in terms of knowledge and skill.

**Instructor's Comments and Lesson Assignment:** Space is provided for your instructor's critique of the lesson, which you can refer to later. Your instructor may also write any specific assignment for the next lesson.

*System suggested by the U.S. Air Force flying clubs

## Reading Assignments for Flight Lessons

You are expected to be prepared for each flight lesson. Our reading assignments will help you understand what is going to happen as well as how and why you need to study diligently BEFORE you go to the airport.

Each flight lesson in this book contains

- Objective
- Text references
- Content
    - ✓ Review items
    - ✓ New items
- Completion standards
- Comments, assignment, and notes

Next to each item in the Content section, we have provided the page number(s) to read in *Commercial Pilot Flight Maneuvers and Practical Test Prep* (FM), the study units/subunits to read in Multi-Engine Add-On Rating Course (MARC), and the section to read, if appropriate, in your airplane's Pilot's Operating Handbook (POH).

Note: The lesson items often refer to MARC content from multiple study units. Because the MARC study unit outlines can only be accessed one at a time through the course interface, we recommend that you complete the entire MARC training program and save the course study unit outlines to your computer for offline use. You may also print the course outlines and place them in a binder for easy reference both during your training program and after your training is complete.

For the new items, you should read the material and attempt to understand the basic concepts. Try to anticipate and visualize the concepts and flight maneuvers. With this basic knowledge, your instructor can expand on the specific and finer points, especially when explaining how a task is done in your specific airplane.

After your flight lesson, task items are fresh in your mind; they will make sense, and you should be able to understand and learn more.

Study review items so you can explain them to your instructor and, at the completion of this syllabus, to your examiner.

After you study, relax and plan a time to begin reading to prepare for the next flight lesson.

## FLIGHT LESSON 1: INTRODUCTION TO MULTI-ENGINE AIRPLANES

### Objective

To introduce the pilot to the multi-engine training airplane, specifically how its operation differs from single-engine airplanes.

### Text References

*Commercial Pilot Flight Maneuvers and Practical Test Prep* (FM)
Multi-Engine Add-On Rating Course (MARC)
Pilot's Operating Handbook (POH)

### Content

1. Preflight briefing
2. New items
   - ☐☐☐ Certificates and documents - FM 68-71
   - ☐☐☐ Airworthiness requirements - FM 72-76
   - ☐☐☐ Operation of systems - FM 106-110; MARC 3, 8.5; POH 1, 7, 8, 9
   - ☐☐☐ Performance and limitations - FM 102-105; MARC 5, 8.4; POH 2, 5, 6
   - ☐☐☐ Preflight inspection - FM 118-122; MARC 6.2; POH 4
   - ☐☐☐ Cockpit management - FM 123-126; MARC 6.3
   - ☐☐☐ Engine starting - FM 127-131; MARC 6.4; POH 4
   - ☐☐☐ Taxiing - FM 132-138; MARC 6.5
   - ☐☐☐ Before-takeoff check - FM 144-147; MARC 6.7; POH 4
   - ☐☐☐ Normal and crosswind takeoff and climb - FM 170-178; MARC 6.8; POH 4
   - ☐☐☐ Traffic patterns - FM 153-159
   - ☐☐☐ Runway incursion and collision avoidance - FM 153-154; MARC 4.9, 4.10
   - ☐☐☐ Normal and crosswind approach and landing - FM 179-196; MARC 6.9; POH 4
   - ☐☐☐ After landing, parking, and securing - FM 336-339; POH 4
   - ☐☐☐ Additional items at CFI's discretion ____________________

     ____________________
3. Postflight critique and preview of next lesson

### Completion Standards

The lesson will have been successfully completed when the pilot demonstrates his/her ability to perform the tasks listed in a multi-engine airplane with minimal assistance from the instructor. The pilot will be able to maintain the desired altitude, ±150 ft.; airspeed, ±15 kt.; and heading, ±20°.

**Instructor's comments:** ____________________

____________________

**Lesson assignment:** ____________________

____________________

**Notes:** ____________________

____________________

____________________

____________________

____________________

____________________

____________________

____________________

____________________

____________________

## FLIGHT LESSON 2: NORMAL MULTI-ENGINE FLIGHT MANEUVERS

### Objective

To review fundamental multi-engine maneuvers and build the pilot's knowledge and proficiency while reducing his/her reliance on instructor guidance.

### Text References

*Commercial Pilot Flight Maneuvers and Practical Test Prep* (FM)
Multi-Engine Add-On Rating Course (MARC)
Pilot's Operating Handbook (POH)

### Content

1. Flight Lesson 1 complete? Yes ____ Copy of lesson placed in pilot's folder? Yes ____
2. Preflight briefing
3. Review items
   - ☐☐☐ Preflight preparation - FM 68-76, 102-110; MARC 3, 8.4, 8.5; POH 1, 2, 5-9
   - ☐☐☐ Preflight inspection - FM 118-122; MARC 6.2; POH 4
   - ☐☐☐ Cockpit management - FM 123-126; MARC 6.3
   - ☐☐☐ Engine starting - FM 127-131; MARC 6.4; POH 4
   - ☐☐☐ Taxiing - FM 132-138; MARC 6.5
   - ☐☐☐ Before-takeoff check - FM 144-147; MARC 6.7; POH 4
   - ☐☐☐ Normal and crosswind takeoff and climb - FM 170-178; MARC 6.8; POH 4
   - ☐☐☐ Traffic patterns - FM 153-159
   - ☐☐☐ Runway incursion and collision avoidance - FM 153-154; MARC 4.9, 4.10, 6.6
   - ☐☐☐ Normal and crosswind approach and landing - FM 179-196; MARC 6.9; POH 4
   - ☐☐☐ After landing, parking, and securing - FM 336-339; POH 4

   ____________________________________________
4. New items
   - ☐☐☐ Maneuvering during slow flight - FM 284-288; MARC 6.13
   - ☐☐☐ Power-off stalls (straight ahead and in turning flight) - FM 289-294; MARC 6.14
   - ☐☐☐ Power-on stalls (straight ahead and in turning flight) - FM 295-300; MARC 6.15
   - ☐☐☐ Accelerated stalls - FM 301-305; MARC 6.16
   - ☐☐☐ Spin awareness - FM 306-308; MARC 8.7; POH 2
   - ☐☐☐ Additional items at CFI's discretion ____________________________________________

   ____________________________________________
5. Postflight critique and preview of next lesson

### Completion Standards

The lesson will have been successfully completed when the pilot demonstrates his/her ability to perform the tasks listed in a multi-engine airplane with minimal assistance from the instructor. The pilot will be able to maintain the desired altitude, ±150 ft.; airspeed, ±15 kt.; and heading, ±20°.

**Instructor's comments:** ____________________________________________

____________________________________________

**Lesson assignment:** ____________________________________________

____________________________________________

**Notes:** ____________________________________________

____________________________________________

____________________________________________

____________________________________________

____________________________________________

____________________________________________

____________________________________________

## FLIGHT LESSON 3: MULTI-ENGINE PERFORMANCE MANEUVERS

### Objective

To enhance the pilot's knowledge of multi-engine operations by exposing him/her to more advanced maneuvers.

### Text References

*Commercial Pilot Flight Maneuvers and Practical Test Prep* (FM)
Multi-Engine Add-On Rating Course (MARC)
Pilot's Operating Handbook (POH)

### Content

1. Flight Lesson 2 complete? Yes ___ Copy of lesson placed in pilot's folder? Yes ___
2. Preflight briefing
3. Review items
   - ☐☐☐ Preflight preparation - FM 68-76, 102-110; MARC 3, 8.4, 8.5; POH 1, 2, 5-9
   - ☐☐☐ Preflight procedures - FM 118-147; MARC 6.2-6.5, 6.7; POH 4
   - ☐☐☐ Normal and crosswind takeoff and climb - FM 170-178; MARC 6.8; POH 4
   - ☐☐☐ Traffic patterns - FM 153-159
   - ☐☐☐ Runway incursion and collision avoidance - FM 153-154; MARC 4.9, 4.10, 6.6
   - ☐☐☐ Maneuvering during slow flight - FM 284-288; MARC 6.13
   - ☐☐☐ Power-off stalls (straight ahead and in turning flight) - FM 289-294; MARC 6.14
   - ☐☐☐ Power-on stalls (straight ahead and in turning flight) - FM 295-300; MARC 6.15
   - ☐☐☐ Accelerated stalls - FM 301-305; MARC 6.16
   - ☐☐☐ Spin awareness - FM 306-308; MARC 8.7; POH 2
   - ☐☐☐ Normal and crosswind approach and landing - FM 179-196; MARC 6.9; POH 4
   - ☐☐☐ After landing, parking, and securing - FM 336-339; POH 4

   ___________________________
4. New items
   - ☐☐☐ Steep turns - FM 238-242; MARC 6.12
   - ☐☐☐ Short-field takeoff and maximum performance climb - FM 208-214; MARC 6.10
   - ☐☐☐ Short-field approach and landing - FM 215-221; MARC 6.11
   - ☐☐☐ Go-around/rejected landing - FM 231-235; POH 4
   - ☐☐☐ High altitude operations - FM 325-334; POH 7, 8
   - ☐☐☐ Additional items at CFI's discretion ___________________________

   ___________________________
5. Postflight critique and preview of next lesson

### Completion Standards

The lesson will have been successfully completed when the pilot demonstrates his/her ability to perform the tasks listed in a multi-engine airplane with minimal assistance from the instructor. The pilot will be able to maintain the desired altitude, ±150 ft.; airspeed, ±15 kt.; and heading, ±20°.

**Instructor's comments:** ___________________________

___________________________

**Lesson assignment:** ___________________________

___________________________

**Notes:** ___________________________

___________________________

___________________________

___________________________

___________________________

___________________________

## FLIGHT LESSON 4: INSTRUMENT FLYING SKILLS

### Objective

To review and enhance the pilot's ability to perform instrument pilot tasks, specifically in a multi-engine airplane.

### Text References

*Commercial Pilot Flight Maneuvers and Practical Test Prep* (FM)
Multi-Engine Add-On Rating Course (MARC)
Pilot's Operating Handbook (POH)

### Content

1. Flight Lesson 3 complete? Yes ____ Copy of lesson placed in pilot's folder? Yes ____
2. Preflight briefing
3. Review items
   - ☐☐☐ Preflight preparation - FM 68-76, 102-110; MARC 3, 8.4, 8.5; POH 1, 2, 5-9
   - ☐☐☐ Preflight procedures - FM 118-147; MARC 6.2-6.5, 6.7; POH 4
   - ☐☐☐ Short-field takeoff and maximum performance climb - FM 208-214; MARC 6.10; POH 4
   - ☐☐☐ Steep turns - FM 238-242; MARC 6.12
   - ☐☐☐ Go-around/rejected landing - FM 231-235; POH 4
   - ☐☐☐ High altitude operations - FM 325-334; POH 7, 8
   - ☐☐☐ Short-field approach and landing - FM 215-221; MARC 6.11; POH 4

   ____________________________________________
4. New items (all items by instrument reference only)
   - ☐☐☐ Attitude instrument flying - MARC 9
   - ☐☐☐ Steep turns - CFI
   - ☐☐☐ Recovery from unusual attitudes - MARC 9.6
   - ☐☐☐ VOR/Localizer/GPS/ADF orientation and tracking - FM 275-276
   - ☐☐☐ VOR/Localizer/GPS/ADF holding* - MARC 9.7
   - ☐☐☐ VOR/Localizer/ILS/GPS/ADF approach(es)* - MARC 9.8
   - ☐☐☐ Additional items at CFI's discretion ____________________________________________

   ____________________________________________

   *These tasks need only be completed if the pilot currently holds an instrument rating or seeks to add one during the multi-engine practical test. These tasks should be completed using whatever equipment is installed and available in the training airplane.
5. Postflight critique and preview of next lesson

### Completion Standards

The lesson will have been successfully completed when the pilot demonstrates his/her ability to perform the tasks listed in a multi-engine airplane with minimal assistance from the instructor. The pilot will be able to demonstrate the appropriate level of proficiency per the Commercial Pilot Practical Test Standards. Instrument reference tasks should be performed at the proficiency level specified in the Instrument Rating Practical Test Standards.

**Instructor's comments:** ____________________________________________

____________________________________________

**Lesson assignment:** ____________________________________________

____________________________________________

**Notes:** ____________________________________________

____________________________________________

____________________________________________

____________________________________________

____________________________________________

## FLIGHT LESSON 5: ABNORMAL MULTI-ENGINE FLIGHT MANEUVERS

### Objective

To expose the pilot to various failures, especially engine failures, and the operational considerations surrounding those failures when operating a multi-engine airplane.

### Text References

*Commercial Pilot Flight Maneuvers and Practical Test Prep* (FM)
Multi-Engine Add-On Rating Course (MARC)
Pilot's Operating Handbook (POH)

### Content

1. Flight Lesson 4 complete? Yes ____ Copy of lesson placed in pilot's folder? Yes ____
2. Preflight briefing
3. Review items
   - ☐☐☐ Attitude instrument flying (IR) - MARC 9
   - ☐☐☐ Recovery from unusual attitudes (IR) – MARC 9.6
   - ☐☐☐ VOR/Localizer/GPS/ADF orientation and tracking (IR) - FM 275-276
   - ☐☐☐ VOR/Localizer/GPS/ADF holding (IR)* - MARC 9.7
   - ☐☐☐ VOR/Localizer/ILS/GPS/ADF approach(es) (IR)* - MARC 9.8

   *These tasks need only be completed if the pilot currently holds an instrument rating or seeks to add one during the multi-engine practical test. These tasks should be completed using whatever equipment is installed and available in the training airplane.

4. New items
   - ☐☐☐ Systems and equipment malfunctions - FM 321-322; MARC 7.6; POH 3, 9
   - ☐☐☐ Maneuvering with one engine inoperative - MARC 7.7, 8.6
   - ☐☐☐ $V_{MC}$ demonstration - MARC 7.8
   - ☐☐☐ Partial panel instrument flying skills (IR) - CFI
     - ☐☐☐ Loss of primary flight instrument indicators (IR) - CFI
     - ☐☐☐ Constant airspeed climbs and descents (IR) - CFI
     - ☐☐☐ Magnetic compass turns (IR) - CFI
     - ☐☐☐ Timed turns (IR) - CFI
     - ☐☐☐ Recovery from unusual attitudes (IR) - CFI
   - ☐☐☐ Additional items at CFI's discretion ________________________________________

   ________________________________________________________________

   IR means instrument references only

5. Postflight critique and preview of next lesson

### Completion Standards

The lesson will have been successfully completed when the pilot demonstrates his/her ability to perform the tasks listed in a multi-engine airplane with minimal assistance from the instructor. The pilot will be able to demonstrate the appropriate level of proficiency per the Commercial Pilot Practical Test Standards. Instrument reference tasks should be performed at the proficiency level specified in the Instrument Rating Practical Test Standards.

**Instructor's comments:** ________________________________________________

________________________________________________________________

**Lesson assignment:** ________________________________________________

________________________________________________________________

**Notes:** ________________________________________________________

________________________________________________________________

________________________________________________________________

________________________________________________________________

## FLIGHT LESSON 6: EMERGENCY OPERATIONS

### Objective

To expose the pilot to various emergency situations that can occur during flight in a multi-engine airplane. Emphasis will be placed on emergency prevention, recognition, and resolution.

### Text References

*Commercial Pilot Flight Maneuvers and Practical Test Prep* (FM)
Multi-Engine Add-On Rating Course (MARC)
Pilot's Opinion Handbook (POH)

### Content

1. Flight Lesson 5 complete? Yes ____ Copy of lesson placed in pilot's folder? Yes ____
2. Preflight briefing
3. Review items
   - ☐☐☐ Systems and equipment malfunctions - FM 321-322; MARC 7.6; POH 3, 9
   - ☐☐☐ Maneuvering with one engine inoperative - MARC 7.7, 8.6
   - ☐☐☐ $V_{MC}$ demonstration - MARC 7.8
   - ☐☐☐ Full panel instrument flying skills (IR) - MARC 9
   - ☐☐☐ Partial panel instrument flying skills (IR) - CFI

   ________________________________________

4. New items
   - ☐☐☐ Engine failure during takeoff before $V_{MC}$ - MARC 7.3
   - ☐☐☐ Engine failure during takeoff after lift-off - MARC 7.4
   - ☐☐☐ Emergency descent - MARC 7.2
   - ☐☐☐ Engine failure during flight by reference to instruments (IR)* - MARC 7.9
   - ☐☐☐ Instrument approaches – one engine inoperative (IR)* - MARC 7.10
   - ☐☐☐ Additional items at CFI's discretion ________________________________________

   ________________________________________

   *These tasks need only be completed if the pilot currently holds an instrument rating or seeks to add one during the multi-engine practical test. These tasks should be completed using whatever equipment is installed and available in the training airplane.

5. Postflight critique and preview of next lesson

### Completion Standards

The lesson will have been successfully completed when the pilot demonstrates his/her ability to perform the tasks listed in a multi-engine airplane with minimal assistance from the instructor. The pilot will be able to demonstrate the appropriate level of proficiency per the Commercial Pilot Practical Test Standards.

**Instructor's comments:** ________________________________________

________________________________________

**Lesson assignment:** ________________________________________

________________________________________

**Notes:** ________________________________________

________________________________________

________________________________________

________________________________________

________________________________________

________________________________________

________________________________________

________________________________________

## FLIGHT LESSON 7: MULTI-ENGINE DAY CROSS-COUNTRY

### Objective

To expose the pilot to real-world multi-engine operations by conducting a daytime cross-country flight.

### Text References

*Commercial Pilot Flight Maneuvers and Practical Test Prep* (FM)
Multi-Engine Add-On Rating Course (MARC)
Pilot's Operating Handbook (POH)

### Content

1. Flight Lesson 6 complete? Yes ____ Copy of lesson placed in pilot's folder? Yes ____
2. Preflight briefing
3. Review items
   - ☐☐☐ Performance and weight and balance limitations - FM 102-105; MARC 5; POH 2, 5, 6
   - ☐☐☐ Systems and equipment malfunctions - FM 321-322; MARC 7.6; POH 3, 9
   - ☐☐☐ Maneuvering with one engine inoperative - MARC 7.7, 8.6
   - ☐☐☐ $V_{MC}$ demonstration - MARC 7.8
   - ☐☐☐ Engine failure during takeoff before $V_{MC}$ - MARC 7.3
   - ☐☐☐ Engine failure during takeoff after lift-off - MARC 7.4
   - ☐☐☐ Emergency descent - FM 311-313; MARC 7.2; POH 3
   - ☐☐☐ Full panel instrument flying skills (IR) - MARC 9
   - ☐☐☐ Partial panel instrument flying skills (IR) - CFI
   - ☐☐☐ Engine failure during flight by reference to instruments (IR)* - MARC 7.9
   - ☐☐☐ VOR/Localizer/GPS/ADF holding (IR)* - MARC 9.7
   - ☐☐☐ Instrument approaches – all engines operating (IR)* - MARC 9.8
   - ☐☐☐ Instrument approaches – one engine inoperative (IR)* - MARC 7.10

   *These tasks need only be completed if the pilot currently holds an instrument rating or seeks to add one during the multi-engine practical test. These tasks should be completed using whatever equipment is installed and available in the training airplane.

   ________________________________________
4. New items
   - ☐☐☐ Cross-country flight planning - FM 92-94
   - ☐☐☐ Weather briefing - FM 77-91
   - ☐☐☐ Pilotage and dead reckoning - FM 270-274
   - ☐☐☐ Lost procedures - FM 280-282
   - ☐☐☐ Diversion to an alternate airport - FM 277-279
   - ☐☐☐ Airspace/ATC operations/procedures - FM 95-101, 150-152
   - ☐☐☐ Airport, runway, and taxiway signs, markings, and lighting - FM 160-167
   - ☐☐☐ Aeromedical factors - FM 111-116
   - ☐☐☐ Additional items at CFI's discretion ________________________________________

   ________________________________________________________________
5. Postflight critique and preview of next lesson

NOTE: This lesson may be completed under VFR or IFR at the discretion of the flight instructor and/or chief flight instructor, as appropriate.

### Completion Standards

The lesson will have been successfully completed when the pilot demonstrates his/her ability to perform the tasks listed in a multi-engine airplane with minimal assistance from the instructor. The flight must be no less than 2 hours in duration and cover a distance greater than 100 nautical miles from the point of departure. The pilot will be able to demonstrate the appropriate level of proficiency per the Commercial Pilot Practical Test Standards.

**Instructor's comments:** ________________________________________

**Lesson assignment:** ________________________________________

**Notes:** ________________________________________

## FLIGHT LESSON 8: MULTI-ENGINE NIGHT CROSS-COUNTRY

### Objective

To expose the pilot to real-world multi-engine operations by conducting a nighttime cross-country flight.

### Text References

*Commercial Pilot Flight Maneuvers and Practical Test Prep* (FM)
Multi-Engine Add-On Rating Course (MARC)
Pilot's Operating Handbook (POH)

### Content

1. Flight Lesson 7 complete? Yes ____ Copy of lesson placed in pilot's folder? Yes ____
2. Preflight briefing
3. Review items

☐☐☐ Systems and equipment malfunctions - FM 321-322; MARC 7.6; POH 3, 9
☐☐☐ Maneuvering with one engine inoperative - MARC 7.7, 8.6
☐☐☐ $V_{MC}$ demonstration - MARC 7.8
☐☐☐ Engine failure during takeoff before $V_{MC}$ - MARC 7.3
☐☐☐ Engine failure during takeoff after lift-off - MARC 7.4
☐☐☐ Emergency descent - FM 311-313; MARC 7.2; POH 3
☐☐☐ Full panel instrument flying skills (IR) - MARC 9
☐☐☐ Partial panel instrument flying skills (IR) - CFI
☐☐☐ Engine failure during flight by reference to instruments (IR)* - MARC 7.9
☐☐☐ VOR/Localizer/GPS/ADF holding (IR)* - MARC 9.7
☐☐☐ Instrument approaches – all engines operating (IR)* - MARC 9.8
☐☐☐ Instrument approaches – one engine inoperative (IR)* - MARC 7.10
☐☐☐ Cross-country flight planning - FM 92-94
☐☐☐ Performance and weight and balance limitations - FM 102-105; MARC 5; POH 2, 5, 6
☐☐☐ Weather briefing - FM 77-91
☐☐☐ Pilotage and dead reckoning - FM 270-274
☐☐☐ Lost procedures - FM 280-282
☐☐☐ Diversion to an alternate airport - FM 277-279
☐☐☐ Airspace/ATC operations/procedures - FM 95-101, 150-152
☐☐☐ Airport, runway, and taxiway signs, markings, and lighting - FM 160-167

*These tasks need only be completed if the pilot currently holds an instrument rating or seeks to add one during the multi-engine practical test. These tasks should be completed using whatever equipment is installed and available in the training airplane.

4. New items

☐☐☐ Aeromedical factors associated with night flight – CFI
☐☐☐ Required pilot/airplane equipment for night flight – CFI
☐☐☐ Additional items at CFI's discretion ____________________

5. Postflight critique and preview of next lesson

NOTE: This lesson may be completed under VFR or IFR at the discretion of the flight instructor and/or chief flight instructor, as appropriate.

### Completion Standards

The lesson will have been successfully completed when the pilot demonstrates his/her ability to perform the tasks listed in a multi-engine airplane with minimal assistance from the instructor. The flight must be no less than 2 hours in duration and cover a distance greater than 100 nautical miles from the point of departure. The pilot will be able to demonstrate the appropriate level of proficiency per the Commercial Pilot Practical Test Standards.

**Instructor's comments:** ____________________

**Lesson assignment:** ____________________

**Notes:** ____________________

# FLIGHT LESSON 9: MANEUVERS REVIEW

## Objective

To prepare the pilot for the end-of-course check and the Commercial Pilot Multi-Engine Land Add-On Practical Test.

## Text References

*Commercial Pilot Flight Maneuvers and Practical Test Prep* (FM)
Multi-Engine Add-On Rating Course (MARC)
Pilot's Operating Handbook (POH)

## Content

1. Flight Lesson 8 complete? Yes ___ Copy of lesson placed in pilot's folder? Yes ___
2. Preflight briefing
3. Review items

- ☐☐☐ Preflight preparation - FM 68-76, 102-110; MARC 3, 8.4, 8.5; POH 1, 2, 5-9
- ☐☐☐ Preflight inspection - FM 118-122; MARC 6.2; POH 4
- ☐☐☐ Cockpit management - FM 123-126; MARC 6.3
- ☐☐☐ Engine starting - FM 127-131; MARC 6.4; POH 4
- ☐☐☐ Taxiing - FM 132-138; MARC 6.5
- ☐☐☐ Before-takeoff check - FM 144-147; MARC 6.7; POH 4
- ☐☐☐ Normal and crosswind takeoff and climb - FM 170-178; MARC 6.8; POH 4
- ☐☐☐ Traffic patterns - FM 153-159
- ☐☐☐ Runway incursion and collision avoidance - FM 153-154; MARC 4.9, 4.10, 6.6
- ☐☐☐ Normal and crosswind approach and landing - FM 179-196; MARC 6.9; POH 4
- ☐☐☐ After landing, parking, and securing - FM 336-339; POH 4
- ☐☐☐ Maneuvering during slow flight - FM 284-288; MARC 6.13
- ☐☐☐ Power-off stalls (straight ahead and in turning flight) - FM 289-294; MARC 6.14
- ☐☐☐ Power-on stalls (straight ahead and in turning flight) - FM 295-300; MARC 6.15
- ☐☐☐ Accelerated stalls - FM 301-305; MARC 6.16
- ☐☐☐ Spin awareness - FM 306-308; MARC 8.7; POH 2
- ☐☐☐ Steep turns - FM 238-242; MARC 6.12
- ☐☐☐ Short-field takeoff and maximum performance climb - FM 208-214; MARC 6.10; POH 4
- ☐☐☐ Short-field approach and landing - FM 215-221; MARC 6.11; POH 4
- ☐☐☐ Go-around/rejected landing - FM 231-235; POH 4
- ☐☐☐ High altitude operations - FM 325-334; POH 7, 8
- ☐☐☐ Attitude instrument flying - MARC 9
- ☐☐☐ Steep turns (IR) - CFI
- ☐☐☐ Recovery from unusual attitudes (IR) - MARC 9.6
- ☐☐☐ VOR/Localizer/GPS/ADF orientation and tracking (IR)* - FM 275-276
- ☐☐☐ VOR/Localizer/GPS/ADF holding (IR)* - MARC 9.7
- ☐☐☐ VOR/Localizer/ILS/GPS/ADF approach(es) (IR)* - MARC 9.8
- ☐☐☐ Partial panel instrument flying skills (IR) - CFI
  - ☐☐☐ Loss of primary flight instrument indicators (IR) - CFI
  - ☐☐☐ Constant airspeed climbs and descents (IR) - CFI
  - ☐☐☐ Magnetic compass turns (IR) - CFI
  - ☐☐☐ Timed turns (IR) - CFI
  - ☐☐☐ Recovery from unusual attitudes (IR) - CFI
- ☐☐☐ Systems and equipment malfunctions - FM 321-322; MARC 7.6; POH 3, 9
- ☐☐☐ Maneuvering with one engine inoperative - MARC 7.7, 8.6
- ☐☐☐ $V_{MC}$ demonstration - MARC 7.8
- ☐☐☐ Engine failure during takeoff before $V_{MC}$ - MARC 7.3
- ☐☐☐ Engine failure during takeoff after lift-off - MARC 7.4
- ☐☐☐ Emergency descent - FM 311-313; MARC 7.2; POH 3
- ☐☐☐ Engine failure during flight by reference to instruments (IR)* - MARC 7.9
- ☐☐☐ Instrument approaches - one engine inoperative (IR)* - MARC 7.10
- ☐☐☐ Additional items at CFI's discretion ________________

*These tasks need only be completed if the pilot currently holds an instrument rating or seeks to add one during the multi-engine practical test. These tasks should be completed using whatever equipment is installed and available in the training airplane.

4. Postflight critique and preview of next lesson

## Completion Standards

The lesson will have been successfully completed when the pilot demonstrates his/her ability to perform the tasks listed in a multi-engine airplane with minimal assistance from the instructor. The pilot will be able to demonstrate the appropriate level of proficiency per the Commercial Pilot Practical Test Standards.

**Instructor's comments:** ________________

**Lesson assignment:** ________________

**Notes:** ________________

## FLIGHT LESSON 10: END-OF-COURSE CHECK

### Objective

To ensure that the pilot meets all applicable standards for the Commercial Pilot Multi-Engine Land Add-On Practical Test.

### Text References

*Commercial Pilot Flight Maneuvers and Practical Test Prep* (FM)
Multi-Engine Add-On Rating Course (MARC)
Pilot's Operating Handbook (POH)

### Content

1. Flight Lesson 9 complete? Yes ____ Copy of lesson placed in pilot's folder? Yes ____
2. Preflight briefing
3. Review items

☐☐☐ Preflight preparation - FM 68-76, 102-110; MARC 3, 8.4, 8.5; POH 1, 2, 5-9
☐☐☐ Preflight inspection - FM 118-122; MARC 6.2; POH 4
☐☐☐ Cockpit management - FM 123-126; MARC 6.3
☐☐☐ Engine starting - FM 127-131; MARC 6.4; POH 4
☐☐☐ Taxiing - FM 132-138; MARC 6.5
☐☐☐ Before-takeoff check - FM 144-147; MARC 6.7; POH 4
☐☐☐ Normal and crosswind takeoff and climb - FM 170-178; MARC 6.8; POH 4
☐☐☐ Traffic patterns - FM 153-159
☐☐☐ Runway incursion and collision avoidance - FM 153-154; MARC 4.9, 4.10, 6.6
☐☐☐ Normal and crosswind approach and landing - FM 179-196; MARC 6.9; POH 4
☐☐☐ After landing, parking, and securing - FM 336-339; POH 4
☐☐☐ Maneuvering during slow flight - FM 284-288; MARC 6.13
☐☐☐ Power-off stalls (straight ahead and in turning flight) - FM 289-294; MARC 6.14
☐☐☐ Power-on stalls (straight ahead and in turning flight) - FM 295-300; MARC 6.15
☐☐☐ Accelerated stalls - FM 301-305; MARC 6.16
☐☐☐ Spin awareness - FM 306-308; POH 2
☐☐☐ Steep turns - FM 238-242; MARC 6.12
☐☐☐ Short-field takeoff and maximum performance climb - FM 208-214; MARC 6.10; POH 4
☐☐☐ Short-field approach and landing - FM 215-221; MARC 6.11; POH 4
☐☐☐ Go-around/rejected landing - FM 231-235; POH 4
☐☐☐ High altitude operations - FM 325-334; POH 7, 8
☐☐☐ Attitude instrument flying - MARC 9
☐☐☐ Steep turns (IR) - CFI
☐☐☐ Recovery from unusual attitudes (IR) - MARC 9.6
☐☐☐ VOR/Localizer/GPS/ADF orientation and tracking (IR)* - FM 275-276
☐☐☐ VOR/Localizer/GPS/ADF holding (IR)* - MARC 9.7
☐☐☐ VOR/Localizer/ILS/GPS/ADF approach(es) (IR)* - MARC 9.8
☐☐☐ Partial panel instrument flying skills (IR) - CFI
- ☐☐☐ Loss of primary flight instrument indicators (IR) - CFI
- ☐☐☐ Constant airspeed climbs and descents (IR) - CFI
- ☐☐☐ Magnetic compass turns (IR) - CFI
- ☐☐☐ Timed turns (IR) - CFI
- ☐☐☐ Recovery from unusual attitudes (IR) CFI

☐☐☐ Systems and equipment malfunctions - FM 321-322; MARC 7.6; POH 3, 9
☐☐☐ Maneuvering with one engine inoperative - MARC 7.7, 8.6
☐☐☐ $V_{MC}$ demonstration - MARC 7.8
☐☐☐ Engine failure during takeoff before $V_{MC}$ - MARC 7.3
☐☐☐ Engine failure during takeoff after lift-off - MARC 7.4
☐☐☐ Emergency descent - FM 311-313; MARC 7.2; POH 3
☐☐☐ Engine failure during flight by reference to instruments (IR)* - MARC 7.9
☐☐☐ Instrument approaches - one engine inoperative (IR)* - MARC 7.10
☐☐☐ Additional items at CFI's discretion ____________
____________

*These tasks need only be completed if the pilot currently holds an instrument rating or seeks to add one during the multi-engine practical test. These tasks should be completed using whatever equipment is installed and available in the training airplane.

4. Flight Lesson 10 complete? Yes ____ Copy of lesson and graduation certificate placed in pilot's folder? Yes ____
5. Postflight critique

### Completion Standards

The lesson and the course of training will have been successfully completed when the pilot demonstrates his/her ability to perform the tasks listed in a multi-engine airplane and demonstrate the appropriate level of proficiency per the Commercial Pilot Practical Test Standards. If additional instruction is necessary, the chief flight instructor will assign the additional training. If the flight is satisfactory, the chief flight instructor will complete the pilot's training records and issue a graduation certificate.

**Instructor's comments:** ____________

**Notes:** ____________

# APPENDIX A
# KNOWLEDGE TESTS AND FIGURES - AIRPLANE SINGLE-ENGINE LAND

For a copy of the answer key to all *Commercial Pilot Syllabus* Stage Tests, please email a request for the CPSYL Stage Test Answer Key to the Gleim Aviation Department at aviation@gleim.com.

**Table of Contents – Single Engine**

# STAGE ONE KNOWLEDGE TEST - SINGLE-ENGINE

**1.** (Refer to Figure 15 on page 148.) GIVEN:

Airport pressure altitude. . . . . . . . . . . . . . . 2,000 ft
Airport temperature . . . . . . . . . . . . . . . . . . 20°C
Cruise pressure altitude . . . . . . . . . . . . . 10,000 ft
Cruise temperature . . . . . . . . . . . . . . . . . . . 0°C

What will be the fuel, time, and distance required to climb to cruise altitude under the given conditions?

A — 5 gallons, 9 minutes, 13 NM.
B — 6 gallons, 11 minutes, 16 NM.
C — 7 gallons, 12 minutes, 18 NM.

**2.** (Refer to Figure 33 on page 152.) GIVEN:

Weight. . . . . . . . . . . . . . . . . . . . . . . . . . 3,700 lb
Pressure altitude. . . . . . . . . . . . . . . . . . . 22,000 ft
Temperature . . . . . . . . . . . . . . . . . . . . . . –10°C

What is the maximum rate of climb under the given conditions?

A — 305 ft/min.
B — 320 ft/min.
C — 384 ft/min.

**3.** GIVEN:

Pressure altitude. . . . . . . . . . . . . . . . . . . . 6,000 ft
True air temperature . . . . . . . . . . . . . . . . . . +30°F

From the conditions given, the approximate density altitude is

A — 9,000 feet.
B — 5,500 feet.
C — 5,000 feet.

**4.** (Refer to Figure 32 on page 151.) GIVEN:

Temperature . . . . . . . . . . . . . . . . . . . . . . . . 30°F
Pressure altitude. . . . . . . . . . . . . . . . . . . . 6,000 ft
Weight. . . . . . . . . . . . . . . . . . . . . . . . . . 3,300 lb
Headwind . . . . . . . . . . . . . . . . . . . . . . . . . 20 kts

What is the total takeoff distance over a 50-foot obstacle?

A — 1,100 feet.
B — 1,300 feet.
C — 1,500 feet.

**5.** (Refer to Figure 3 on page 145.) If an airplane glides at an angle of attack of 10°, how much altitude will it lose in 1 mile?

A — 240 feet.
B — 480 feet.
C — 960 feet.

**6.** The uncontrolled firing of the fuel/air charge in advance of normal spark ignition is known as

A — instantaneous combustion.
B — detonation.
C — pre-ignition.

**7.** GIVEN:

Total weight. . . . . . . . . . . . . . . . . . . . . . 4,137 lb
CG location. . . . . . . . . . . . . . . . . . . . . Station 67.8
Fuel consumption. . . . . . . . . . . . . . . . . . . 13.7 GPH
Fuel CG. . . . . . . . . . . . . . . . . . . . . . . Station 68.0

After 1 hour 30 minutes of flight time, the CG would be located at station

A — 67.79
B — 68.79
C — 70.78

**8.** During the transition from straight-and-level flight to a climb, the angle of attack is increased and lift

A — is momentarily decreased.
B — remains the same.
C — is momentarily increased.

**9.** (Refer to Figure 51 on page 155.) While clearing an active runway you are most likely clear of the ILS critical area when you pass which symbol?

A — Top red.
B — Middle yellow.
C — Bottom yellow.

**10.** (Refer to Figure 64 on page 155.) You are holding short for an intersection departure on Runway 8 with the sign in front of you. After turning onto the runway you should

A — Turn right.
B — Turn left.
C — Insufficient information is given.

**11.** When operating an airplane for the purpose of takeoff or landing within Class D airspace under special VFR, what minimum distance from clouds and what visibility are required?

A — Remain clear of clouds, and the ground visibility must be at least 1 SM.
B — 500 feet beneath clouds, and the ground visibility must be at least 1 SM.
C — Remain clear of clouds, and the flight visibility must be at least 1 NM.

**12.** Applying carburetor heat will

A — not affect the mixture.
B — lean the fuel/air mixture.
C — enrich the fuel/air mixture.

**13.** On a wing, the force of lift acts perpendicular to and the force of drag acts parallel to the

A — chord line.
B — flightpath.
C — longitudinal axis.

**14.** Longitudinal stability involves the motion of the airplane controlled by its

A — rudder.
B — elevator.
C — ailerons.

**15.** The ratio between the total airload imposed on the wing and the gross weight of an aircraft in flight is known as

A — load factor and directly affects stall speed.
B — aspect load and directly affects stall speed.
C — load factor and has no relation with stall speed.

**16.** What will occur if no leaning is made with the mixture control as the flight altitude increases?

A — The volume of air entering the carburetor decreases and the amount of fuel decreases.
B — The density of air entering the carburetor decreases and the amount of fuel increases.
C — The density of air entering the carburetor decreases and the amount of fuel remains constant.

**17.** Notification to the NTSB is required when there has been substantial damage

A — which requires repairs to landing gear.
B — to an engine caused by engine failure in flight.
C — which adversely affects structural strength or flight characteristics.

**18.** The operator of an aircraft that has been involved in an incident is required to submit a report to the nearest field office of the NTSB

A — within 7 days.
B — within 10 days.
C — only if requested to do so.

**19.** The carriage of passengers for hire by a commercial pilot is

A — not authorized in limited category aircraft.
B — not authorized in utility category airplane.
C — authorized in restricted category aircraft.

**20.** When operating a U.S.-registered civil aircraft, which document is required by regulation to be available in the aircraft?

A — A manufacturer's Operations Manual.
B — A current, approved Airplane Flight Manual.
C — An Owner's Manual.

**21.** Which is required to operate an aircraft towing an advertising banner?

A — Approval from ATC to operate in Class E airspace.
B — A certificate of waiver issued by the Administrator.
C — A safety link at each end of the towline which has a breaking strength not less than 80 percent of the aircraft's gross weight.

**22.** Which is required equipment for powered aircraft during VFR night flights?

A — Flashlight with red lens, if the flight is for hire.
B — An electric landing light, if the flight is for hire.
C — Sensitive altimeter adjustable for barometric pressure.

**23.** Which is true with respect to formation flights? Formation flights are

A — not authorized, except by arrangement with the pilot in command of each aircraft.
B — not authorized, unless the pilot in command of each aircraft is trained and found competent in formation.
C — authorized when carrying passengers for hire, with prior arrangement with the pilot in command of each aircraft in the formation.

**24.** The required preflight action relative to weather reports and fuel requirements is applicable to

A — any flight conducted for compensation or hire.
B — any flight not in the vicinity of an airport.
C — IFR flights only.

**25.** VFR cruising altitudes are required to be maintained when flying

A — at 3,000 feet or more AGL; based on true course.
B — more than 3,000 feet AGL; based on magnetic course.
C — at 3,000 feet or more above MSL; based on magnetic heading.

# STAGE TWO KNOWLEDGE TEST - SINGLE-ENGINE

**1.** An airplane descends to an airport under the following conditions:

Cruising altitude . . . . . . . . . . . . . . . . . 6,500 ft
Airport elevation . . . . . . . . . . . . . . . . . . . 700 ft
Descends to. . . . . . . . . . . . . . . . . . 800 ft AGL
Rate of descent . . . . . . . . . . . . . . . . 500 ft/min
Average true airspeed . . . . . . . . . . . . . 110 kts
True course . . . . . . . . . . . . . . . . . . . . . . 335°
Average wind velocity . . . . . . . . . 060° at 15 kts
Variation . . . . . . . . . . . . . . . . . . . . . . . . 3°W
Deviation . . . . . . . . . . . . . . . . . . . . . . . . +2°
Average fuel consumption . . . . . . . . . 8.5 gal/hr

Determine the approximate time, compass heading, distance, and fuel consumed during the descent.

A — 10 minutes, 348°, 18 NM, 1.4 gallons.
B — 10 minutes, 355°, 17 NM, 2.4 gallons.
C — 12 minutes, 346°, 18 NM, 1.6 gallons.

**2.** Inbound on the 040 radial, a pilot selects the 055 radial, turns 15° to the left, and notes the time. While maintaining a constant heading, the pilot notes the time for the CDI to center is 15 minutes. Based on this information, the ETE to the station is

A — 8 minutes.
B — 15 minutes.
C — 30 minutes.

**3.** Inbound on the 315 radial, a pilot selects the 320 radial, turns 5° to the left, and notes the time. While maintaining a constant heading, the pilot notes the time for the CDI to center is 12 minutes. The ETE to the station is

A — 10 minutes.
B — 12 minutes.
C — 24 minutes.

**4.** Inbound on the 190 radial, a pilot selects the 195 radial, turns 5° to the left, and notes the time. While maintaining a constant heading, the pilot notes the time for the CDI to center is 10 minutes. The ETE to the station is

A — 10 minutes.
B — 15 minutes.
C — 20 minutes.

**5.** An airplane descends to an airport under the following conditions:

Cruising altitude . . . . . . . . . . . . . . . . 10,500 ft
Airport elevation . . . . . . . . . . . . . . . . . 1,700 ft
Descends to. . . . . . . . . . . . . . . . . 1,000 ft AGL
Rate of descent . . . . . . . . . . . . . . . . 600 ft/min
Average true airspeed . . . . . . . . . . . . . 135 kts
True course . . . . . . . . . . . . . . . . . . . . . . 263°
Average wind velocity . . . . . . . . . 330° at 30 kts
Variation . . . . . . . . . . . . . . . . . . . . . . . . 7°E
Deviation . . . . . . . . . . . . . . . . . . . . . . . . +3°
Average fuel consumption . . . . . . . . 11.5 gal/hr

Determine the approximate time, compass heading, distance, and fuel consumed during the descent.

A — 9 minutes, 274°, 26 NM, 2.8 gallons.
B — 13 minutes, 274°, 28 NM, 2.5 gallons.
C — 13 minutes, 271°, 26 NM, 2.5 gallons.

**6.** The magnetic heading is 315° and the ADF shows a relative bearing of 140°. The magnetic bearing FROM the radiobeacon would be

A — 095°.
B — 175°.
C — 275°.

**7.** (Refer to Figure 17 on page 149.) Which is true regarding illustration 4, if the present heading is maintained? The airplane will

A — cross the 060 radial at a 15° angle.
B — intercept the 240 radial at a 30° angle.
C — cross the 180 radial at a 75° angle.

**8.** The remarks section of the Aviation Routine Weather Report (METAR) contains the following coded information. What does it mean?

RMK FZDZB42 WSHFT 30 FROPA

A — Freezing drizzle with cloud bases below 4,200 feet.
B — Freezing drizzle below 4,200 feet and wind shear.
C — Wind shift at three zero due to frontal passage.

**9.** To best determine observed weather conditions between weather reporting stations, the pilot should refer to

A — pilot reports.
B — Area Forecasts.
C — prognostic charts.

**10.** Which is true about homing when using ADF during crosswind conditions? Homing

A — to a radio station results in a curved path that leads to the station.
B — is a practical navigation method for flying both to and from a radio station.
C — to a radio station requires that the ADF have an automatically or manually rotatable azimuth.

**11.** Hypoxia is the result of which of these conditions?

A — Excessive oxygen in the bloodstream.
B — Insufficient oxygen reaching the brain.
C — Excessive carbon dioxide in the bloodstream.

**12.** The basic drive for a pilot to demonstrate the "right stuff" can have an adverse effect on safety, by

A — a total disregard for any alternative course of action.
B — generating tendencies that lead to practices that are dangerous, often illegal, and may lead to a mishap.
C — allowing events, or the situation, to control his or her actions.

**13.** Which is true regarding preheating an aircraft during cold weather operations?

A — The cabin area as well as the engine should be preheated.
B — The cabin area should not be preheated with portable heaters.
C — Hot air should be blown directly at the engine through the air intakes.

**14.** Which would decrease the stability of an air mass?

A — Warming from below.
B — Cooling from below.
C — Decrease in water vapor.

**15.** Convective currents are most active on warm summer afternoons when winds are

A — light.
B — moderate.
C — strong.

**16.** One of the most dangerous features of mountain waves is the turbulent areas in and

A — below rotor clouds.
B — above rotor clouds.
C — below lenticular clouds.

**17.** Which correctly describes the purpose of convective SIGMETs (WST)?

A — They consist of an hourly observation of tornadoes, significant thunderstorm activity, and large hailstone activity.
B — They contain both an observation and a forecast of all thunderstorm and hailstone activity. The forecast is valid for 1 hour only.
C — They consist of either an observation and a forecast or just a forecast for tornadoes, significant thunderstorm activity, or hail greater than or equal to 3/4 inch in diameter.

**18.** The section of the Aviation Area Forecast (FA) entitled VFR Clouds and Weather contains a summary of

A — forecast sky cover, cloud tops, visibility, and obstructions to vision along specific routes.
B — only those weather systems producing liquid or frozen precipitation, fog, thunderstorms, or IFR ceilings.
C — sky condition, cloud heights, visibility, obstructions to vision, precipitation, and sustained surface winds of 20 knots or greater.

**19.** Which conditions are favorable for the formation of a surface based temperature inversion?

A — Clear, cool nights with calm or light wind.
B — Area of unstable air rapidly transferring heat from the surface.
C — Broad areas of cumulus clouds with smooth, level bases at the same altitude.

**20.** Which would most likely result in hyperventilation?

A — Insufficient oxygen.
B — Excessive carbon monoxide.
C — Insufficient carbon dioxide.

**21.** The most severe weather conditions, such as destructive winds, heavy hail, and tornadoes, are generally associated with

A — slow-moving warm fronts which slope above the tropopause.
B — squall lines.
C — fast-moving occluded fronts.

**22.** When taxiing during strong quartering tailwinds, which aileron positions should be used?

A — Neutral.
B — Aileron up on the side from which the wind is blowing.
C — Aileron down on the side from which the wind is blowing.

**23.** Which is true regarding the development of convective circulation?

A — Cool air must sink to force the warm air upward.
B — Warm air is less dense and rises on its own accord.
C — Warmer air covers a larger surface area than the cool air; therefore, the warmer air is less dense and rises.

**24.** During the life cycle of a thunderstorm, which stage is characterized predominately by downdrafts?

A — Mature.
B — Developing.
C — Dissipating.

**25.** During an approach, the most important and most easily recognized means of being alerted to possible wind shear is monitoring the

A — amount of trim required to relieve control pressures.
B — heading changes necessary to remain on the runway centerline.
C — power and vertical velocity required to remain on the proper glidepath.

# END-OF-COURSE KNOWLEDGE TEST - SINGLE-ENGINE

**1.** (Refer to Figure 20 on page 150.) Which instrument shows the aircraft in a position where a 180° turn would result in the aircraft intercepting the 150 radial at a 30° angle?

A — 2.
B — 3.
C — 4.

**2.** (Refer to Figure 2 on page 145.) Select the correct statement regarding stall speeds. The airplane will stall

A — 10 knots higher in a power-on, 60° bank, with gear and flaps up, than with gear and flaps down.
B — 25 knots lower in a power-off, flaps-up, 60° bank, than in a power-off, flaps-down, wings-level configuration.
C — 10 knots higher in a 45° bank, power-on stall, than in a wings-level stall with flaps up.

**3.** (Refer to Figure 33 on page 152.) GIVEN:

| | |
|---|---|
| Weight | 4,000 lb |
| Pressure altitude | 5,000 ft |
| Temperature | 30°C |

What is the maximum rate of climb under the given conditions?

A — 655 ft/min.
B — 702 ft/min.
C — 774 ft/min.

**4.** An airplane departs an airport under the following conditions:

| | |
|---|---|
| Airport elevation | 1000 ft |
| Cruise altitude | 9,500 ft |
| Rate of climb | 500 ft/min |
| Average true airspeed | 135 kts |
| True course | 215° |
| Average wind velocity | 290° at 20 kts |
| Variation | 3° W |
| Deviation | –2° |
| Average fuel consumption | 13 gal/hr |

Determine the approximate time, compass heading, distance, and fuel consumed during the climb.

A — 14 minutes, 234°, 26 NM, 3.9 gallons.
B — 17 minutes, 224°, 36 NM, 3.7 gallons.
C — 17 minutes, 242°, 31 NM, 3.5 gallons.

**5.** (Refer to Figure 35 on page 154.) GIVEN:

| | |
|---|---|
| Temperature | 70°F |
| Pressure altitude | Sea level |
| Weight | 3,400 lb |
| Headwind | 16 kts |

Determine the approximate ground roll.

A — 689 feet.
B — 716 feet.
C — 1,275 feet.

**6.** (Refer to Figure 32 on page 151.) GIVEN:

| | |
|---|---|
| Temperature | 50°F |
| Pressure altitude | 2,000 ft |
| Weight | 2,700 lb |
| Wind | Calm |

What is the total takeoff distance over a 50-foot obstacle?

A — 800 feet.
B — 650 feet.
C — 1,050 feet.

**7.** (Refer to Figure 15 on page 148.) GIVEN:

| | |
|---|---|
| Airport pressure altitude | 4,000 ft |
| Airport temperature | 12°C |
| Cruise pressure altitude | 9,000 ft |
| Cruise temperature | – 4°C |

What will be the distance required to climb to cruise altitude under the given conditions?

A — 6 miles.
B — 8.5 miles.
C — 11 miles.

**8.** (Refer to Figure 34 on page 153.) GIVEN:

| | |
|---|---|
| Pressure altitude | 6,000 ft |
| Temperature | +3°C |
| Power | 2,200 RPM – 22" MP |
| Usable fuel available | 465 lb |

What is the maximum available flight time under the conditions stated?

A — 6 hours 27 minutes.
B — 6 hours 39 minutes.
C — 6 hours 56 minutes.

**9.** (Refer to Figure 8 on page 147.) GIVEN:

| | |
|---|---|
| Fuel quantity | 47 gal |
| Power-cruise (lean) | 55 percent |

Approximately how much flight time would be available with a night VFR fuel reserve remaining?

A — 3 hours 8 minutes.
B — 3 hours 22 minutes.
C — 3 hours 43 minutes.

**10.** If an airplane is consuming 95 pounds of fuel per hour at a cruising altitude of 6,500 feet and the groundspeed is 173 knots, how much fuel is required to travel 450 NM?

A — 248 pounds.
B — 265 pounds.
C — 284 pounds.

**11.** (Refer to Figure 24 on page 148.) If the time flown between aircraft positions 2 and 3 is 15 minutes, what is the estimated time to the station?

A — 15 minutes.
B — 30 minutes.
C — 60 minutes.

**12.** (Refer to Figure 23 on page 148.) If the time flown between aircraft positions 2 and 3 is 13 minutes, what is the estimated time to the station?

A — 7.8 minutes.
B — 13 minutes.
C — 26 minutes.

**13.** Inbound on the 090 radial, a pilot rotates the OBS 010° to the left, turns 010° to the right, and notes the time. While maintaining a constant heading, the pilot determines that the elapsed time for the CDI to center is 8 minutes. Based on this information, the ETE to the station is

A — 8 minutes.
B — 16 minutes.
C — 24 minutes.

**14.** GIVEN:

| | |
|---|---|
| True course | 105° |
| True heading | 085° |
| True airspeed | 95 kts |
| Groundspeed | 87 kts |

Determine the wind direction and speed.

A — 020° and 32 knots.
B — 030° and 38 knots.
C — 200° and 32 knots.

**15.** (Refer to Figure 11 on page 146.) If the cruise altitude is 7,500 feet, using 64 percent power at 2,500 RPM, what would be the range with 48 gallons of usable fuel?

A — 635 miles.
B — 645 miles.
C — 810 miles.

**16.** The station originating the following METAR observation has a field elevation of 3,500 feet MSL. If the sky cover is one continuous layer, what is the thickness of the cloud layer? (Top of overcast reported at 7,500 feet MSL.)

METAR KHOB 151250Z 17006KT 4SM OVC005
13/11 A2998

A — 2,500 feet.
B — 3,500 feet.
C — 4,000 feet.

**17.** What is meant by the Special METAR weather observation for KBOI?

SPECI KBOI 091854Z 32005KT 1 1/2SM RA BR
OVC007 17/16 A2990 RMK RAB12

A — Rain and fog obscuring two-tenths of the sky; rain began at 1912Z.
B — Rain and mist obstructing visibility; rain began at 1812Z.
C — Rain and overcast at 1,200 feet AGL.

**18.** What weather phenomenon is implied within an area enclosed by small scalloped lines on a U.S. High-Level Significant Weather Prognostic Chart?

A — Cirriform clouds, light to moderate turbulence, and icing.
B — Cumulonimbus clouds, icing, and moderate or greater turbulence.
C — Cumuliform or standing lenticular clouds, moderate to severe turbulence, and icing.

**19.** (Refer to Figure 5 on page 146.) The horizontal dashed line from point C to point E represents the

A — ultimate load factor.
B — positive limit load factor.
C — airspeed range for normal operations.

**20.** Which situation would result in reverse sensing of a VOR receiver?

A — Flying a heading that is reciprocal to the bearing selected on the OBS.
B — Setting the OBS to a bearing that is 90° from the bearing on which the aircraft is located.
C — Failing to change the OBS from the selected inbound course to the outbound course after passing the station.

**21.** Which is true regarding tracking on a desired bearing when using ADF during crosswind conditions?

A — To track outbound, heading corrections should be made away from the ADF pointer.
B — When on the desired track outbound with the proper drift correction established, the ADF pointer will be deflected to the windward side of the tail position.
C — When on the desired track inbound with the proper drift correction established, the ADF pointer will be deflected to the windward side of the nose position.

**22.** (Refer to Figure 20 on page 150.) Which instrument shows the aircraft to be northwest of the VORTAC?

A — 1.
B — 2.
C — 3.

**23.** To scan properly for traffic, a pilot should

A — slowly sweep the field of vision from one side to the other at intervals.
B — concentrate on any peripheral movement detected.
C — use a series of short, regularly spaced eye movements that bring successive areas of the sky into the central visual field.

**24.** Most pilots have fallen prey to dangerous tendencies or behavior problems at some time. Some of these dangerous tendencies or behavior patterns which must be identified and eliminated include:

A — Deficiencies in instrument skills and knowledge of aircraft systems or limitations.
B — Performance deficiencies from human factors such as fatigue, illness, or emotional problems.
C — Peer pressure, get-there-itis, loss of positional or situational awareness, and operating without adequate fuel reserves.

**25.** What are some of the hazardous attitudes dealt with in Aeronautical Decision Making (ADM)?

A — Antiauthority (don't tell me), impulsivity (do something quickly without thinking), macho (I can do it).
B — Risk management, stress management, and risk elements.
C — Poor decision making, situational awareness, and judgment.

**26.** The pilot and passengers are anxious to get to their destination for a business presentation. Level IV thunderstorms are reported to be in a line across their intended route of flight. Which of the following alternatives best illustrates the IMPULSIVITY reaction?

A — They want to hurry and get going, before things get worse.
B — A thunderstorm won't stop them.
C — They can't change the weather, so they might as well go.

**27.** Detonation occurs in a reciprocating aircraft engine when

A — there is an explosive increase of fuel caused by too rich a fuel/air mixture.
B — the spark plugs receive an electrical jolt caused by a short in the wiring.
C — the unburned fuel/air charge in the cylinders is subjected to instantaneous combustion.

**28.** The need to slow an aircraft below $V_A$ is brought about by the following weather phenomenon:

A — High density altitude which increases the indicated stall speed.
B — Turbulence which causes an increase in stall speed.
C — Turbulence which causes a decrease in stall speed.

**29.** An airplane is loaded to a gross weight of 4,800 pounds, with three pieces of luggage in the rear baggage compartment. The CG is located 98 inches aft of datum, which is 1 inch aft of limits. If luggage which weighs 90 pounds is moved from the rear baggage compartment (145 inches aft of datum) to the front compartment (45 inches aft of datum), what is the new CG?

A — 96.13 inches aft of datum.
B — 95.50 inches aft of datum.
C — 99.87 inches aft of datum.

**30.** An aircraft is loaded with a ramp weight of 3,650 pounds and having a CG of 94.0, approximately how much baggage would have to be moved from the rear baggage area at station 180 to the forward baggage area at station 40 in order to move the CG to 92.0?

A — 52.14 pounds.
B — 62.24 pounds.
C — 78.14 pounds.

**31.** When should pilots decline a "land and hold short" (LAHSO) clearance?

A — When it will compromise safety.
B — If runway surface is contaminated.
C — Only when the tower controller concurs.

**32.** When approaching to land at an airport without an operating control tower, in Class G airspace, the pilot should

A — make all turns to the left, unless otherwise indicated.
B — fly a left-hand traffic pattern at 800 feet AGL.
C — enter and fly a traffic pattern at 800 feet AGL.

**33.** What is the maximum indicated airspeed authorized in the airspace underlying Class B airspace?

A — 156 knots.
B — 200 knots.
C — 230 knots.

**34.** If the minimum safe speed for any particular operation is greater than the maximum speed prescribed in 14 CFR Part 91, the

A — operator must have a Memorandum of Agreement (MOA) with the controlling agency.
B — aircraft may be operated at that speed.
C — operator must have a Letter of Agreement with ATC.

**35.** When approaching to land at an airport with an ATC facility, in Class D airspace, the pilot must establish communications prior to

A — 10 NM, up to and including 3,000 feet AGL.
B — 30 SM, and be transponder equipped.
C — 4 NM, up to and including 2,500 feet AGL.

**36.** A coded transponder equipped with altitude reporting equipment is required for

A — Class A, Class B, and Class C airspace areas.
B — all airspace of the 48 contiguous U.S. and the District of Columbia at and above 10,000 feet MSL (including airspace at and below 2,500 feet above the surface).
C — both of the other answers.

**37.** Which is true regarding flight operations in Class A airspace?

A — Aircraft must be equipped with approved distance measuring equipment (DME).
B — Must conduct operations under instrument flight rules.
C — Aircraft must be equipped with an approved ATC transponder.

**38.** At some airports located in Class D airspace where ground visibility is not reported, takeoffs and landings under special VFR are

A — not authorized.
B — authorized by ATC if the flight visibility is at least 1 SM.
C — authorized only if the ground visibility is observed to be at least 3 SM.

**39.** To operate an airplane under SPECIAL VFR (SVFR) within Class D airspace at night, which is required?

A — The pilot must hold an instrument rating, but the airplane need not be equipped for instrument flight, as long as the weather will remain at or above SVFR minimums.
B — The Class D airspace must be specifically designated as a night SVFR area.
C — The pilot must hold an instrument rating, and the airplane must be equipped for instrument flight.

**40.** A proper crosswind landing on a runway requires that, at the moment of touchdown, the

A — direction of motion of the airplane and its lateral axis be perpendicular to the runway.
B — direction of motion of the airplane and its longitudinal axis be parallel to the runway.
C — downwind wing be lowered sufficiently to eliminate the tendency for the airplane to drift.

**41.** Which is true with respect to a high- or low-pressure system?

A — A high-pressure area or ridge is an area of rising air.
B — A low-pressure area or trough is an area of descending air.
C — A high-pressure area or ridge is an area of descending air.

**42.** Which statement is true concerning the effect of the application of carburetor heat?

A — It enriches the fuel/air mixture.
B — It leans the fuel/air mixture.
C — It has no effect on the fuel/air mixture.

**43.** In the Northern Hemisphere, the wind is deflected to the

A — right by Coriolis force.
B — right by surface friction.
C — left by Coriolis force.

**44.** A pilot's most immediate and vital concern in the event of complete engine failure after becoming airborne on takeoff is

A — maintaining a safe airspeed.
B — landing directly into the wind.
C — turning back to the takeoff field.

**45.** Which statement is true about magnetic deviation of a compass? Deviation

A — varies over time as the agonic line shifts.
B — varies for different headings of the same aircraft.
C — is the same for all aircraft in the same locality.

**46.** Name the four fundamentals involved in maneuvering an aircraft.

A — Power, pitch, bank, and trim.
B — Thrust, lift, turns, and glides.
C — Straight-and-level flight, turns, climbs, and descents.

**47.** With respect to advection fog, which statement is true?

A — It is slow to develop and dissipates quite rapidly.
B — It forms almost exclusively at night or near daybreak.
C — It can appear suddenly during day or night, and it is more persistent than radiation fog.

**48.** The angle of attack at which a wing stalls remains constant regardless of

A — weight, dynamic pressure, bank angle, or pitch attitude.
B — dynamic pressure, but varies with weight, bank angle, and pitch attitude.
C — weight and pitch attitude, but varies with dynamic pressure and bank angle.

**49.** To generate the same amount of lift as altitude is increased, an airplane must be flown at

A — the same true airspeed regardless of angle of attack.
B — a lower true airspeed and a greater angle of attack.
C — a higher true airspeed for any given angle of attack.

**50.** While holding the angle of bank constant in a level turn, if the rate of turn is varied the load factor would

A — remain constant regardless of air density and the resultant lift vector.
B — vary depending upon speed and air density provided the resultant lift vector varies proportionately.
C — vary depending upon the resultant lift vector.

**51.** Turbulence that is encountered above 15,000 feet AGL not associated with cumuliform cloudiness, including thunderstorms, should be reported as

A — severe turbulence.
B — clear air turbulence.
C — convective turbulence.

**52.** When turbulence is encountered during the approach to a landing, what action is recommended and for what primary reason?

A — Increase the airspeed slightly above normal approach speed to attain more positive control.
B — Decrease the airspeed slightly below normal approach speed to avoid overstressing the airplane.
C — Increase the airspeed slightly above normal approach speed to penetrate the turbulence as quickly as possible.

**53.** What flight planning information can a pilot derive from Constant Pressure Analysis Charts?

A — Winds and temperatures aloft.
B — Clear air turbulence and icing conditions.
C — Frontal systems and obstructions to vision aloft.

**54.** The Surface Analysis Chart depicts

A — frontal locations and expected movement, pressure centers, cloud coverage, and obstructions to vision at the time of chart transmission.
B — actual frontal positions, pressure patterns, temperature, dewpoint, wind, weather, and obstructions to vision at the valid time of the chart.
C — actual pressure distribution, frontal systems, cloud heights and coverage, temperature, dewpoint, and wind at the time shown on the chart.

**55.** Dashed lines on a Surface Analysis Chart, if depicted, indicate that the pressure gradient is

A — weak.
B — strong.
C — unstable.

**56.** Terminal Aerodrome Forecasts (TAF) are issued how many times a day and cover what period of time?

A — Four times daily and are usually valid for a 24 hour period.
B — Six times daily and are usually valid for a 24 hour period including a 4-hour categorical outlook.
C — Six times daily and are valid for 12 hours including a 6-hour categorical outlook.

**57.** Which provides a graphic display of both VFR and IFR weather?

A — Surface Weather Map.
B — Radar Summary Chart.
C — Weather Depiction Chart.

**58.** What type of In-Flight Weather Advisories provides an en route pilot with information regarding the possibility of moderate icing, moderate turbulence, winds of 30 knots or more at the surface and extensive mountain obscurement?

A — Convective SIGMETs and SIGMETs.
B — Severe Weather Forecast Alerts (AWW) and SIGMETs.
C — AIRMETs and Center Weather Advisories (CWA).

**59.** How should the pilot make a VOR receiver check when the aircraft is located on the designated checkpoint on the airport surface?

A — Set the OBS on 180° plus or minus 4°; the CDI should center with a FROM indication.
B — Set the OBS on the designated radial. The CDI must center within plus or minus 4° of that radial with a FROM indication.
C — With the aircraft headed directly toward the VOR and the OBS set to 000°, the CDI should center within plus or minus 4° of that radial with a TO indication.

**60.** When the CDI needle is centered during an airborne VOR check, the omnibearing selector and the TO/FROM indicator should read

A — within 4° of the selected radial.
B — within 6° of the selected radial.
C — 0° TO, only if you are due south of the VOR.

**61.** The strength and location of the jet stream is normally

A — weaker and farther north in the summer.
B — stronger and farther north in the winter.
C — stronger and farther north in the summer.

**62.** To track outbound on the 180 radial of a VOR station, the recommended procedure is to set the OBS to

A — 360° and make heading corrections toward the CDI needle.
B — 180° and make heading corrections away from the CDI needle.
C — 180° and make heading corrections toward the CDI needle.

**63.** To track inbound on the 215 radial of a VOR station, the recommended procedure is to set the OBS to

A — 215° and make heading corrections toward the CDI needle.
B — 215° and make heading corrections away from the CDI needle.
C — 035° and make heading corrections toward the CDI needle.

**64.** The mixture control can be adjusted, which

A — prevents the fuel/air combination from becoming too rich at higher altitudes.
B — regulates the amount of airflow through the carburetor's venturi.
C — prevents the fuel/air combination from becoming lean as the airplane climbs.

**65.** Which is correct with respect to rate and radius of turn for an airplane flown in a coordinated turn at a constant altitude?

A — For a specific angle of bank and airspeed, the rate and radius of turn will not vary.
B — To maintain a steady rate of turn, the angle of bank must be increased as the airspeed is decreased.
C — The faster the true airspeed, the faster the rate and larger the radius of turn regardless of the angle of bank.

**66.** In aircraft equipped with constant-speed propellers and normally-aspirated engines, which procedure should be used to avoid placing undue stress on the engine components? When power is being

A — decreased, reduce the RPM before reducing the manifold pressure.
B — increased, increase the RPM before increasing the manifold pressure.
C — increased or decreased, the RPM should be adjusted before the manifold pressure.

**67.** What performance is characteristic of flight at maximum lift/drag ratio in a propeller-driven airplane? Maximum

A — gain in altitude over a given distance.
B — range and maximum distance glide.
C — coefficient of lift and minimum coefficient of drag.

**68.** The reason for variations in geometric pitch (twisting) along a propeller blade is that it

A — permits a relatively constant angle of incidence along its length when in cruising flight.
B — prevents the portion of the blade near the hub from stalling during cruising flight.
C — permits a relatively constant angle of attack along its length when in cruising flight.

**69.** To use VHF/DF facilities for assistance in locating your position, you must have an operative VHF

A — transmitter and receiver.
B — transmitter and receiver, and an operative ADF receiver.
C — transmitter and receiver, and an operative VOR receiver.

**70.** While taxiing on the parking ramp, the landing gear, wheel, and tire are damaged by striking ground equipment. What action would be required to comply with NTSB Part 830?

A — An immediate notification must be filed by the operator of the aircraft with the nearest NTSB field office.
B — A report must be filed with the nearest FAA field office within 7 days.
C — No notification or report is required.

**71.** The operator of an aircraft that has been involved in an incident is required to submit a report to the nearest field office of the NTSB

A — within 7 days.
B — within 10 days.
C — only if requested to do so.

**72.** After an annual inspection has been completed and the aircraft has been returned to service, an appropriate notation should be made

A — on the airworthiness certificate.
B — in the aircraft maintenance records.
C — in the FAA-approved flight manual.

**73.** No person may operate an aircraft that has an experimental airworthiness certificate

A — under instrument flight rules (IFR).
B — when carrying property for hire.
C — when carrying persons or property for hire.

**74.** Who is primarily responsible for maintaining an aircraft in an airworthy condition?

A — The lead mechanic responsible for that aircraft.
B — Pilot in command or operator.
C — Owner or operator of the aircraft.

**75.** When operating a U.S.-registered civil aircraft, which document is required by regulation to be available in the aircraft?

A — A manufacturer's Operations Manual.
B — A current, approved Airplane Flight Manual.
C — An Owner's Manual.

**76.** The pilot in command of an aircraft operated under IFR, in controlled airspace, shall report as soon as practical to ATC when

A — climbing or descending to assigned altitudes.
B — experiencing any malfunctions of navigational, approach, or communications equipment, occurring in flight.
C — requested to contact a new controlling facility.

**77.** An airplane is overtaking a helicopter. Which aircraft has the right-of-way?

A — Helicopter; the pilot should expect to be passed on the right.
B — Airplane; the airplane pilot should alter course to the left to pass.
C — Helicopter; the pilot should expect to be passed on the left.

**78.** During a night operation, the pilot of aircraft #1 sees only the green light of aircraft #2. If the aircraft are converging, which pilot has the right-of-way? The pilot of aircraft

A — #2; aircraft #2 is to the left of aircraft #1.
B — #2; aircraft #2 is to the right of aircraft #1.
C — #1; aircraft #1 is to the right of aircraft #2.

**79.** NTSB Part 830 requires an immediate notification as a result of which incident?

A — Engine failure for any reason during flight.
B — Damage to the landing gear as a result of a hard landing.
C — Any required flight crewmember being unable to perform flight duties because of illness.

**80.** Which is true with respect to formation flights? Formation flights are

A — authorized when carrying passengers for hire, with prior arrangement with the pilot in command of each aircraft in the formation.
B — not authorized when visibilities are less than 3 SM.
C — not authorized when carrying passengers for hire.

**81.** What person is directly responsible for the final authority as to the operation of the airplane?

A — Certificate holder.
B — Pilot in command.
C — Airplane owner/operator.

**82.** When is preflight action required, relative to alternatives available, if the planned flight cannot be completed?

A — IFR flights only.
B — Any flight not in the vicinity of an airport.
C — Any flight conducted for hire or compensation.

**83.** Does a commercial pilot certificate have a specific expiration date?

A — No, it is issued without an expiration date.
B — Yes, it expires at the end of the 24th month after the month in which it was issued.
C — No, but commercial privileges expire if a flight review is not satisfactorily completed each 12 months.

**84.** What limitation is imposed on a newly certificated commercial pilot-airplane, if that person does not hold an instrument rating? The carriage of passengers

A — or property for hire on cross-country flights at night is limited to a radius of 50 NM.
B — for hire on cross-country flights is limited to 50 NM for night flights, but not limited for day flights.
C — for hire on cross-country flights in excess of 50 NM, or for hire at night is prohibited.

**85.** A person with a Commercial Pilot certificate may act as pilot in command of an aircraft for compensation or hire, if that person

A — is qualified in accordance with 14 CFR part 61 and with the applicable parts that apply to the operation.
B — is qualified in accordance with 14 CFR part 61 and has passed a pilot competency check given by an authorized check pilot.
C — holds appropriate category, class ratings, and meets the recent flight experience requirements of 14 CFR part 61.

**86.** If an aircraft is not equipped with an electrical or anticollision light system, no person may operate that aircraft

A — after dark.
B — 1 hour after sunset.
C — after sunset to sunrise.

**87.** A pilot convicted of operating a motor vehicle while either intoxicated by, impaired by, or under the influence of alcohol or a drug is required to provide a

A — written report to the FAA Civil Aeromedical Institute (CAMI) within 60 days after the motor vehicle action.
B — written report to the FAA Civil Aviation Security Division (AMC-700) not later than 60 days after the conviction.
C — the notification of the conviction to an FAA Aviation Medical Examiner (AME) not later than 60 days after the motor vehicle action.

**88.** Each required flight crewmember is required to keep his or her shoulder harness fastened

A — during takeoff and landing only when passengers are aboard the aircraft.
B — while the crewmembers are at their stations, unless he or she is unable to perform required duties.
C — during takeoff and landing, unless he or she is unable to perform required duties.

**89.** With regard to the technique required for a crosswind correction on takeoff, a pilot should use

A — aileron pressure into the wind and initiate the lift-off at a normal airspeed in both tailwheel- and nosewheel-type airplanes.
B — right rudder pressure, aileron pressure into the wind, and higher than normal lift-off airspeed in both tricycle- and conventional-gear airplanes.
C — rudder as required to maintain directional control, aileron pressure into the wind, and higher than normal lift-off airspeed in both conventional- and nosewheel-type airplanes.

**90.** Frost covering the upper surface of an airplane wing usually will cause

A — the airplane to stall at an angle of attack that is higher than normal.
B — the airplane to stall at an angle of attack that is lower than normal.
C — drag factors so large that sufficient speed cannot be obtained for takeoff.

**91.** Select the true statement pertaining to the life cycle of a thunderstorm.

A — Updrafts continue to develop throughout the dissipating stage of a thunderstorm.
B — The beginning of rain at the Earth's surface indicates the mature stage of the thunderstorm.
C — The beginning of rain at the Earth's surface indicates the dissipating stage of the thunderstorm.

**92.** The jet stream and associated clear air turbulence can sometimes be visually identified in flight by

A — dust or haze at flight level.
B — long streaks of cirrus clouds.
C — a constant outside air temperature.

**93.** Which is true regarding the presence of alcohol within the human body?

A — A small amount of alcohol increases vision acuity.
B — An increase in altitude decreases the adverse effect of alcohol.
C — Judgment and decision-making abilities can be adversely affected by even small amounts of alcohol.

**94.** During a takeoff made behind a departing large jet airplane, the pilot can minimize the hazard of wingtip vortices by

A — being airborne prior to reaching the jet's flightpath until able to turn clear of its wake.
B — maintaining extra speed on takeoff and climbout.
C — extending the takeoff roll and not rotating until well beyond the jet's rotation point.

**95.** When landing behind a large aircraft, which procedure should be followed for vortex avoidance?

A — Stay above its final approach flightpath all the way to touchdown.
B — Stay below and to one side of its final approach flightpath.
C — Stay well below its final approach flightpath and land at least 2,000 feet behind.

**96.** What is indicated if ice pellets are encountered at 8,000 feet?

A — Freezing rain at higher altitude.
B — You are approaching an area of thunderstorms.
C — You will encounter hail if you continue your flight.

**97.** The Hazardous Inflight Weather Advisory Service (HIWAS) is a broadcast service over selected VORs that provides

A — SIGMETs and AIRMETs at 15 minutes and 45 minutes past the hour for the first hour after issuance.
B — continuous broadcast of inflight weather advisories.
C — SIGMETs, CONVECTIVE SIGMETs and AIRMETs at 15 minutes and 45 minutes past the hour.

**98.** What causes wind?

A — The Earth's rotation.
B — Air mass modification.
C — Pressure differences.

**99.** The minimum vertical wind shear value critical for probable moderate or greater turbulence is

A — 4 knots per 1,000 feet.
B — 6 knots per 1,000 feet.
C — 8 knots per 1,000 feet.

**100.** GIVEN:

Pressure altitude. . . . . . . . . . . . . . . . . . . . 5,000 ft
True air temperature . . . . . . . . . . . . . . . . . . +30°C

From the conditions given, the approximate density altitude is

A — 7,800 feet.
B — 7,200 feet.
C — 9,000 feet.

# FIGURES FOR KNOWLEDGE TESTS - SINGLE-ENGINE

| GROSS WEIGHT 2750 LBS | | ANGLE OF BANK | | | |
|---|---|---|---|---|---|
| | | LEVEL | 30° | 45° | 60° |
| POWER | | GEAR AND FLAPS UP | | | |
| ON | MPH<br>KTS | 62<br>54 | 67<br>58 | 74<br>64 | 88<br>76 |
| OFF | MPH<br>KTS | 75<br>65 | 81<br>70 | 89<br>77 | 106<br>92 |
| | | GEAR AND FLAPS DOWN | | | |
| ON | MPH<br>KTS | 54<br>47 | 58<br>50 | 64<br>56 | 76<br>66 |
| OFF | MPH<br>KTS | 66<br>57 | 71<br>62 | 78<br>68 | 93<br>81 |

Figure 2. – Stall Speeds.

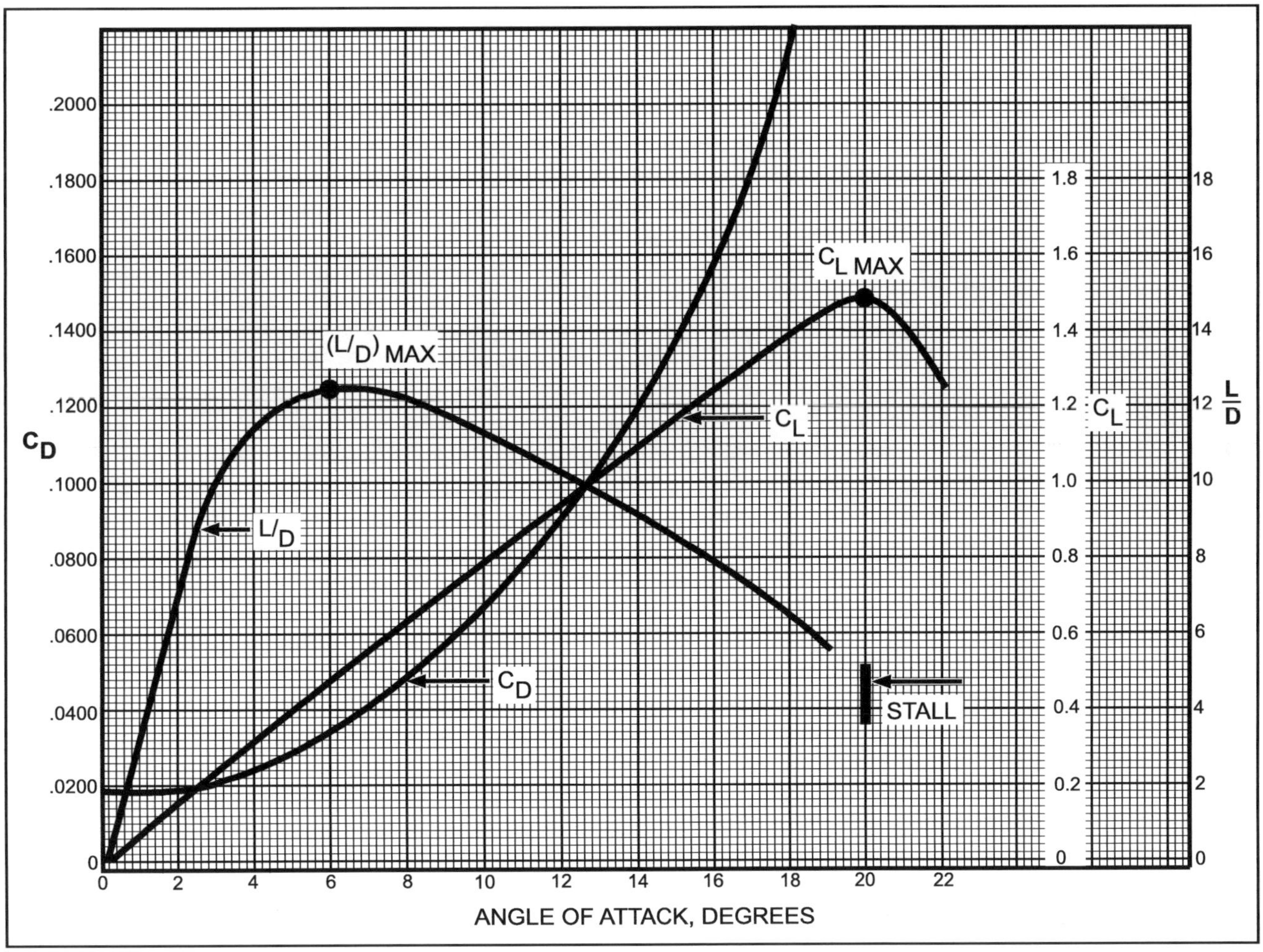


Figure 3. – Angle of Attack, Degrees.

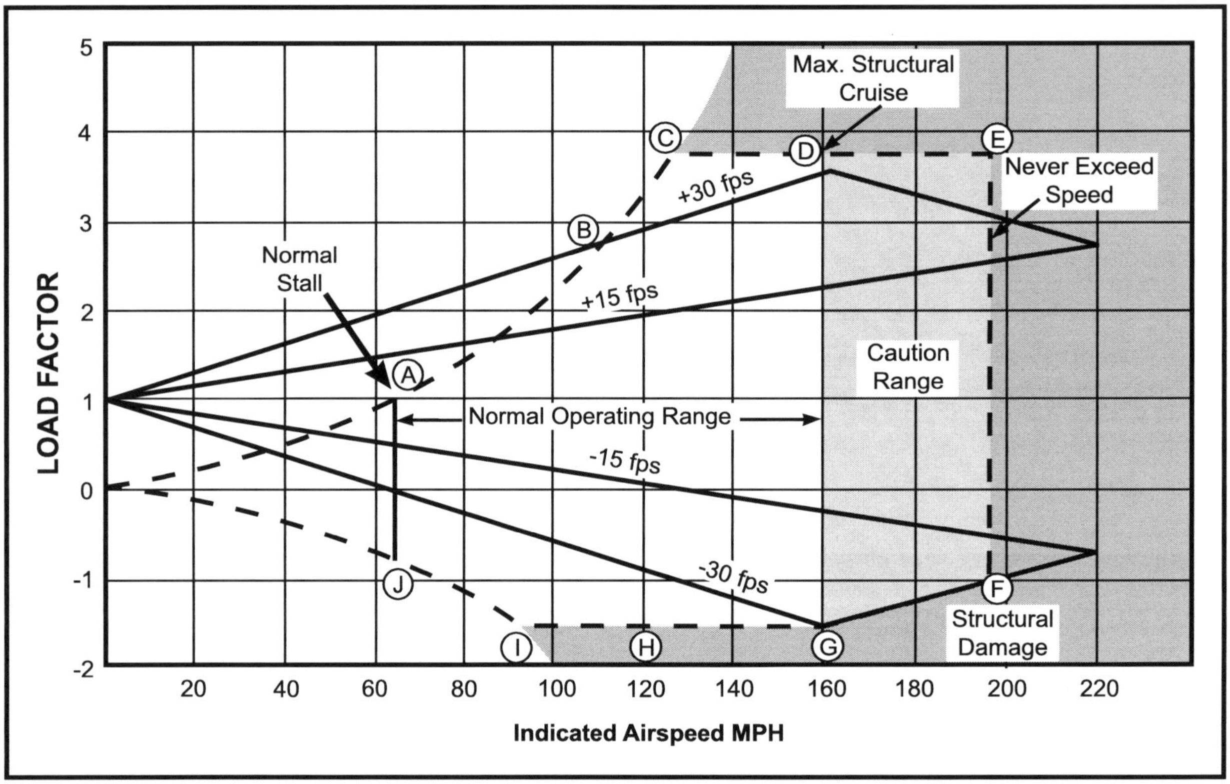


Figure 5. – Velocity vs. G-Loads.

**Gross Weight - 2300 Lbs.**
**Standard Conditions**
**Zero Wind   Lean Mixture**

**NOTE: Maximum cruise is normally limited to 75% power,**

| | | | | | 38 GAL (NO RESERVE) | | 48 GAL (NO RESERVE) | |
|---|---|---|---|---|---|---|---|---|
| ALT. | RPM | % BHP | TAS MPH | GAL/ HOUR | ENDR. HOURS | RANGE MILES | ENDR. HOURS | RANGE MILES |
| 2500 | 2700 | 86 | 134 | 9.7 | 3.9 | 525 | 4.9 | 660 |
| | 2600 | 79 | 129 | 8.6 | 4.4 | 570 | 5.6 | 720 |
| | 2500 | 72 | 123 | 7.8 | 4.9 | 600 | 6.2 | 760 |
| | 2400 | 65 | 117 | 7.2 | 5.3 | 620 | 6.7 | 780 |
| | 2300 | 58 | 111 | 6.7 | 5.7 | 630 | 7.2 | 795 |
| | 2200 | 52 | 103 | 6.3 | 6.1 | 625 | 7.7 | 790 |
| 5000 | 2700 | 82 | 134 | 9.0 | 4.2 | 565 | 5.3 | 710 |
| | 2600 | 75 | 128 | 8.1 | 4.7 | 600 | 5.9 | 760 |
| | 2500 | 68 | 122 | 7.4 | 5.1 | 625 | 6.4 | 790 |
| | 2400 | 61 | 116 | 6.9 | 5.5 | 635 | 6.9 | 805 |
| | 2300 | 55 | 108 | 6.5 | 5.9 | 635 | 7.4 | 805 |
| | 2200 | 49 | 100 | 6.0 | 6.3 | 630 | 7.9 | 795 |
| 7500 | 2700 | 78 | 133 | 8.4 | 4.5 | 600 | 5.7 | 755 |
| | 2600 | 71 | 127 | 7.7 | 4.9 | 625 | 6.2 | 790 |
| | 2500 | 64 | 121 | 7.1 | 5.3 | 645 | 6.7 | 810 |
| | 2400 | 58 | 113 | 6.7 | 5.7 | 645 | 7.2 | 820 |
| | 2300 | 52 | 105 | 6.2 | 6.1 | 640 | 7.7 | 810 |
| 10,000 | 2650 | 70 | 129 | 7.6 | 5.0 | 640 | 6.3 | 810 |
| | 2600 | 67 | 125 | 7.3 | 5.2 | 650 | 6.5 | 820 |
| | 2500 | 61 | 118 | 6.9 | 5.5 | 655 | 7.0 | 830 |
| | 2400 | 55 | 110 | 6.4 | 5.9 | 650 | 7.5 | 825 |
| | 2300 | 49 | 100 | 6.0 | 6.3 | 635 | 8.0 | 800 |

Figure 11. – Cruise and Range Performance.

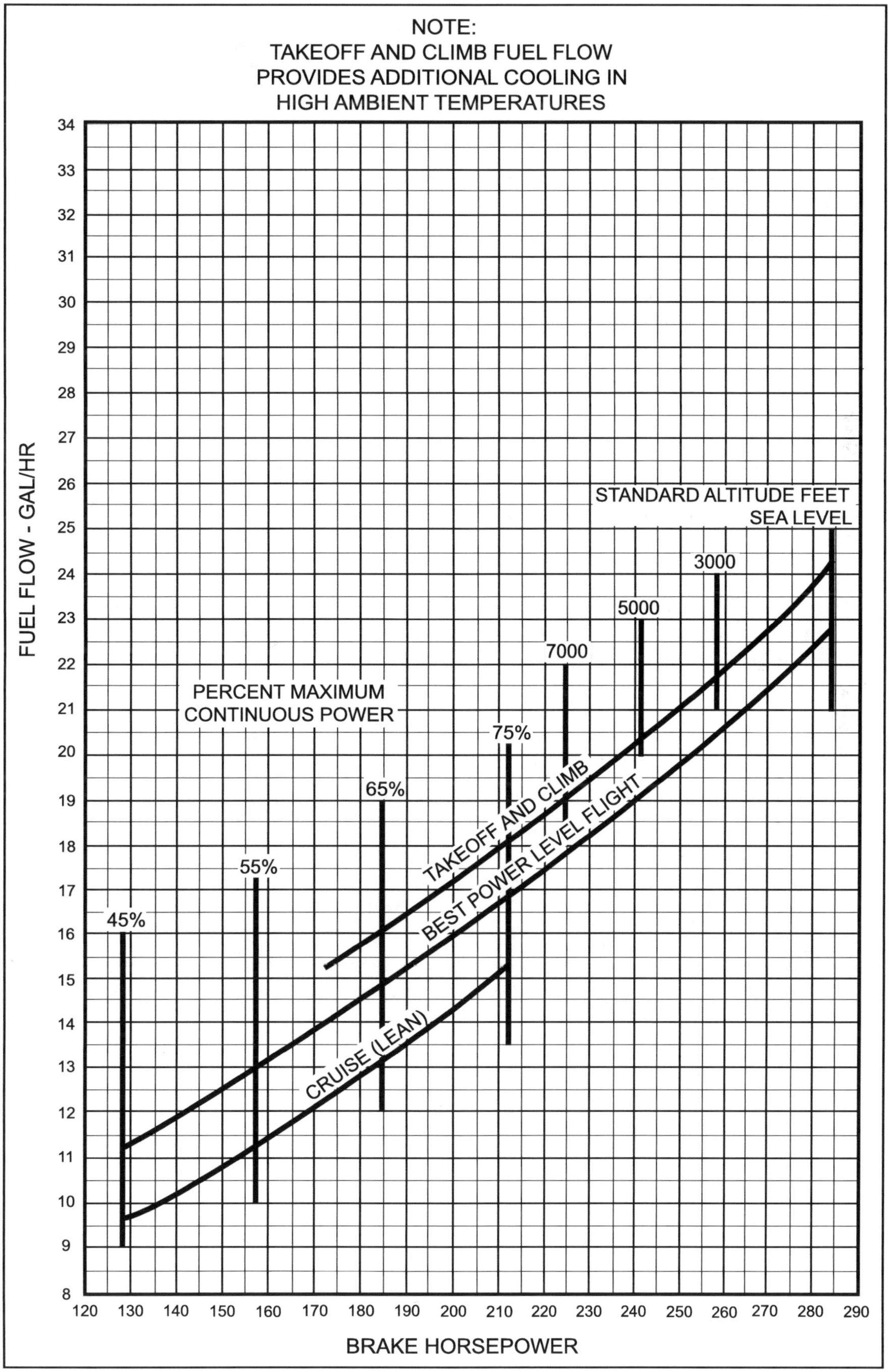


Figure 8. – Fuel Consumption vs. Brake Horsepower.

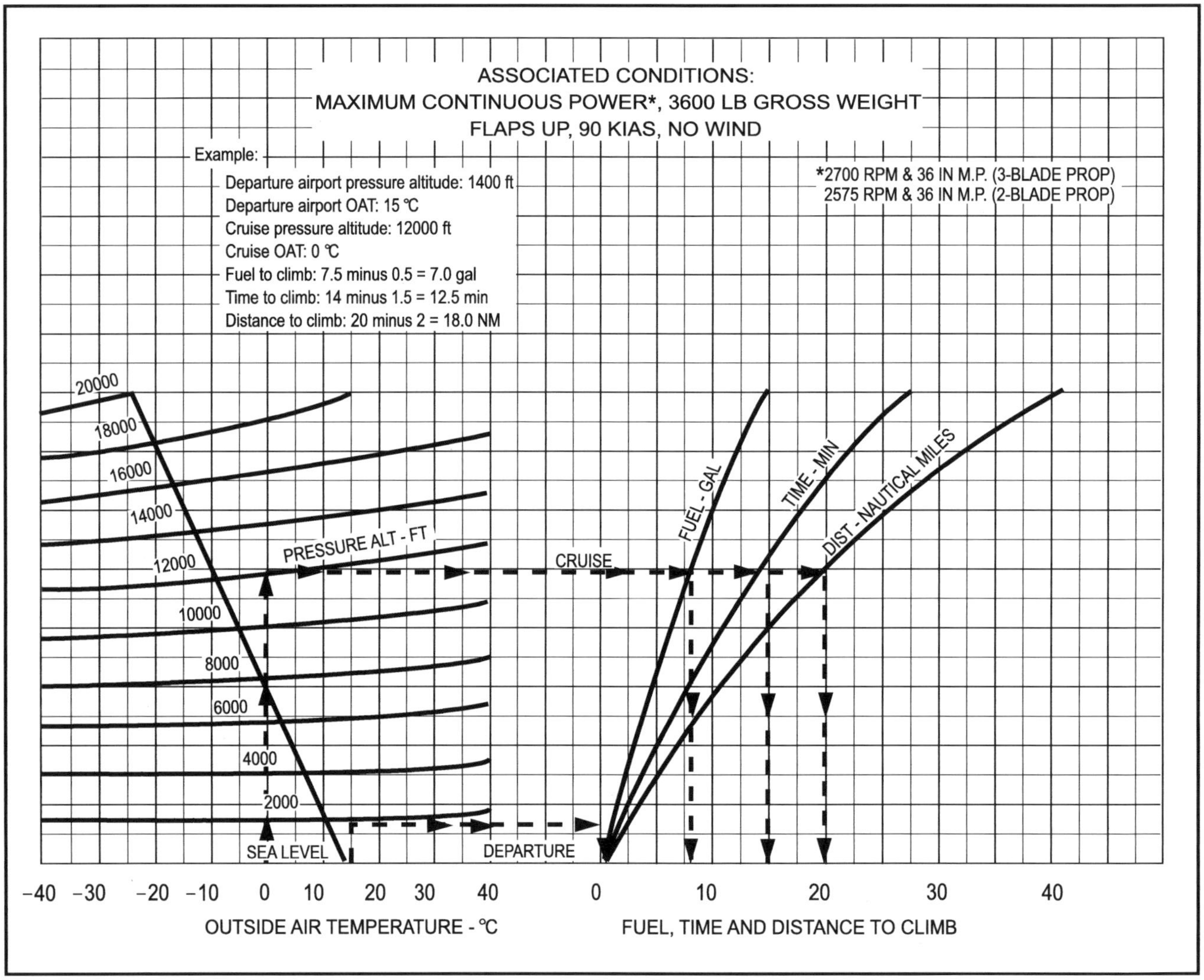


Figure 15. – Fuel, Time, and Distance to Climb.

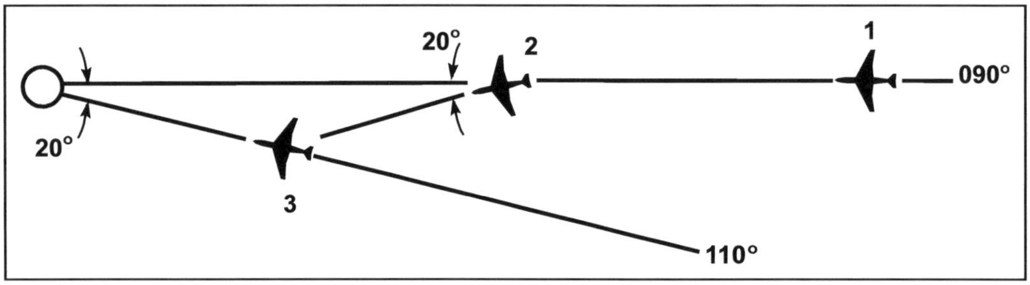


Figure 23. – Isosceles Triangle.

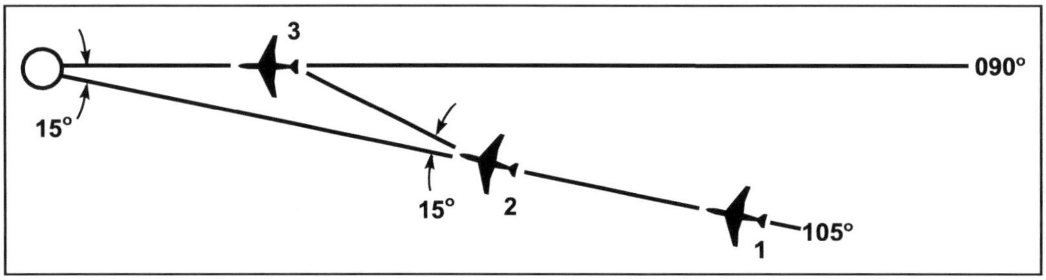


Figure 24. – Isosceles Triangle.

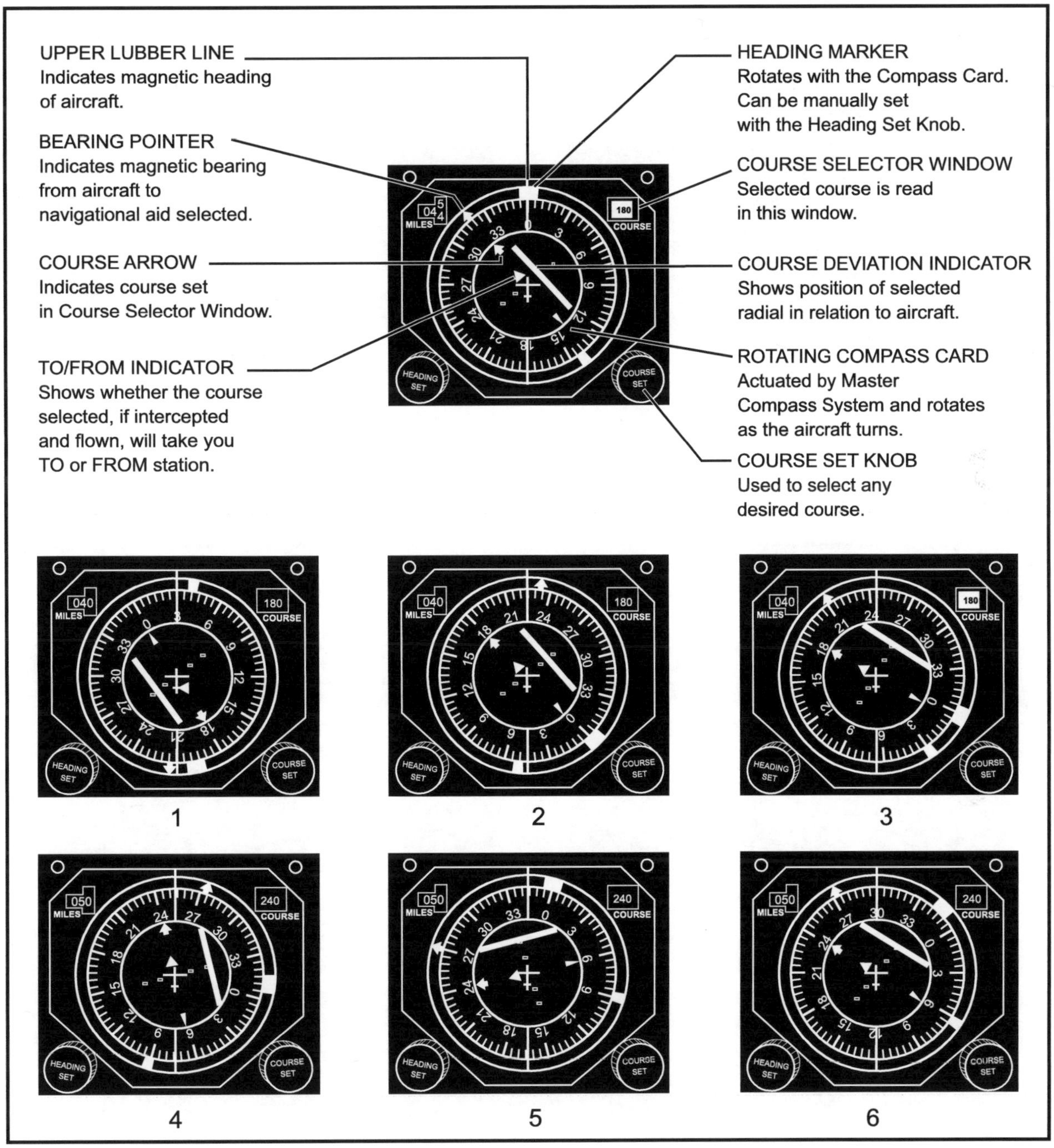


Figure 17. – Horizontal Situation Indicator (HSI).

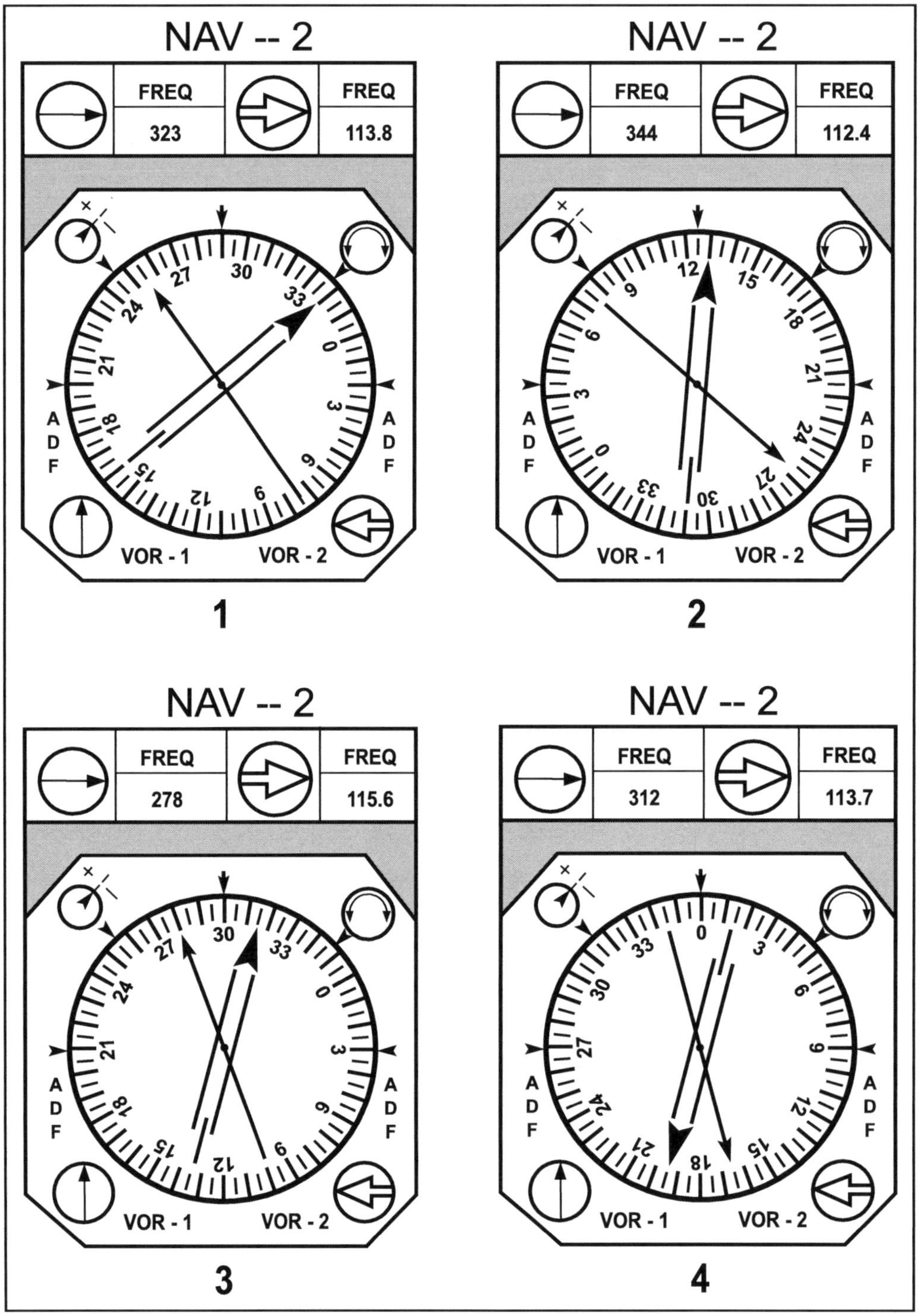

Figure 20. – Radio Magnetic Indicator (RMI).

ASSOCIATED CONDITIONS:

| | |
|---|---|
| POWER | TAKEOFF POWER SET BEFORE BRAKE RELEASE |
| FLAPS | 20* |
| RUNWAY | PAVED, LEVEL, DRY SURFACE |
| TAKEOFF SPEED | IAS AS TABULATED |

NOTE: GROUND ROLL IS APPROX. 73% OF TOTAL TAKEOFF DISTANCE OVER A 50 FT OBSTACLE

EXAMPLE:

| | |
|---|---|
| OAT | 75 F |
| PRESSURE ALTITUDE | 4000 FT |
| TAKEOFF WEIGHT | 3100 LB |
| HEADWIND | 20 KNOTS |
| TOTAL TAKEOFF DISTANCE OVER A 50 FT OBSTACLE | 1350 FT |
| GROUND ROLL (73% OF 1350) | 986 FT |
| IAS TAKEOFF SPEED | |
| LIFT–OFF | 74 MPH |
| AT 50 FT | 74 MPH |

| WEIGHT POUNDS | IAS TAKEOFF SPEED (ASSUMES ZERO INSTR. ERROR) | | | |
|---|---|---|---|---|
| | LIFT–OFF | | 50 FEET | |
| | MPH | KNOTS | MPH | KNOTS |
| 3400 | 77 | 67 | 77 | 67 |
| 3200 | 75 | 65 | 75 | 65 |
| 3000 | 72 | 63 | 72 | 63 |
| 2800 | 69 | 60 | 69 | 60 |
| 2600 | 66 | 57 | 66 | 57 |
| 2400 | 63 | 55 | 63 | 55 |

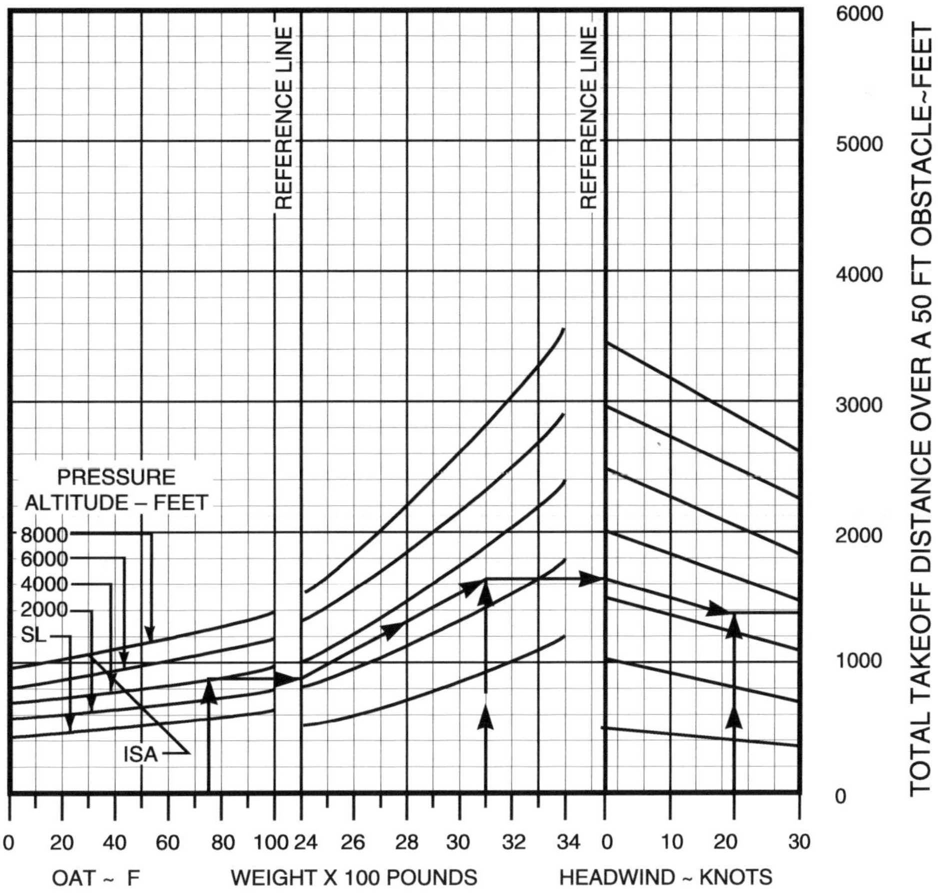


Figure 32. – Obstacle Take-off Chart.

CONDITIONS:
Flaps Up
Gear Up
2600 RPM
Cowl Flaps Open

| PRESS ALT | MP | PPH |
|---|---|---|
| S.L. TO 17,000 | 35 | 162 |
| 18,000 | 34 | 156 |
| 20,000 | 32 | 144 |
| 22,000 | 30 | 132 |
| 24,000 | 28 | 120 |

| WEIGHT LBS | PRESS ALT FT | CLIMB SPEED KIAS | RATE OF CLIMB - FPM | | | |
|---|---|---|---|---|---|---|
| | | | -20 °C | 0 °C | 20 °C | 40 °C |
| 4000 | S.L. | 100 | 1170 | 1035 | 895 | 755 |
| | 4000 | 100 | 1080 | 940 | 800 | 655 |
| | 8000 | 100 | 980 | 840 | 695 | 555 |
| | 12,000 | 100 | 870 | 730 | 590 | --- |
| | 16,000 | 100 | 740 | 605 | 470 | --- |
| | 20,000 | 99 | 485 | 355 | --- | --- |
| | 24,000 | 97 | 190 | 70 | --- | --- |
| 3700 | S.L. | 99 | 1310 | 1165 | 1020 | 875 |
| | 4000 | 99 | 1215 | 1070 | 925 | 775 |
| | 8000 | 99 | 1115 | 965 | 815 | 670 |
| | 12,000 | 99 | 1000 | 855 | 710 | --- |
| | 16,000 | 99 | 865 | 730 | 590 | --- |
| | 20,000 | 97 | 600 | 470 | --- | --- |
| | 24,000 | 95 | 295 | 170 | --- | --- |
| 3400 | S.L. | 97 | 1465 | 1320 | 1165 | 1015 |
| | 4000 | 97 | 1370 | 1220 | 1065 | 910 |
| | 8000 | 97 | 1265 | 1110 | 955 | 795 |
| | 12,000 | 97 | 1150 | 995 | 845 | --- |
| | 16,000 | 97 | 1010 | 865 | 725 | --- |
| | 20,000 | 96 | 730 | 595 | --- | --- |
| | 24,000 | 94 | 405 | 275 | --- | --- |

Figure 33. – Maximum Rate-of-Climb Chart.

# PRESSURE ALTITUDE 6,000 FEET

CONDITIONS:
Recommended Lean Mixture
3800 Pounds
Cowl Flaps Closed

| | | 20 °C BELOW STANDARD TEMP -17 °C | | | STANDARD TEMPERATURE 3 °C | | | 20 °C ABOVE STANDARD TEMP 23 °C | | |
|---|---|---|---|---|---|---|---|---|---|---|
| RPM | MP | % BHP | KTAS | PPH | % BHP | KTAS | PPH | % BHP | KTAS | PPH |
| 2550 | 24 | --- | --- | --- | 78 | 173 | 97 | 75 | 174 | 94 |
| | 23 | 76 | 167 | 96 | 74 | 169 | 92 | 71 | 171 | 89 |
| | 22 | 72 | 164 | 90 | 69 | 166 | 87 | 67 | 167 | 84 |
| | 21 | 68 | 160 | 85 | 65 | 162 | 82 | 63 | 163 | 80 |
| 2500 | 24 | 78 | 169 | 98 | 75 | 171 | 95 | 73 | 172 | 91 |
| | 23 | 74 | 166 | 93 | 71 | 167 | 90 | 69 | 169 | 87 |
| | 22 | 70 | 162 | 88 | 67 | 164 | 85 | 65 | 165 | 82 |
| | 21 | 66 | 158 | 83 | 63 | 160 | 80 | 61 | 160 | 77 |
| 2400 | 24 | 73 | 165 | 91 | 70 | 166 | 88 | 68 | 167 | 85 |
| | 23 | 69 | 161 | 87 | 67 | 163 | 84 | 64 | 164 | 81 |
| | 22 | 65 | 158 | 82 | 63 | 159 | 79 | 61 | 160 | 77 |
| | 21 | 61 | 154 | 77 | 59 | 155 | 75 | 57 | 155 | 73 |
| 2300 | 24 | 68 | 161 | 86 | 66 | 162 | 83 | 64 | 163 | 80 |
| | 23 | 65 | 158 | 82 | 62 | 159 | 79 | 60 | 159 | 76 |
| | 22 | 61 | 154 | 77 | 59 | 155 | 75 | 57 | 155 | 72 |
| | 21 | 57 | 150 | 73 | 55 | 150 | 71 | 53 | 150 | 68 |
| 2200 | 24 | 63 | 156 | 80 | 61 | 157 | 77 | 59 | 158 | 75 |
| | 23 | 60 | 152 | 76 | 58 | 153 | 73 | 56 | 154 | 71 |
| | 22 | 57 | 149 | 72 | 54 | 149 | 70 | 53 | 149 | 67 |
| | 21 | 53 | 144 | 68 | 51 | 144 | 66 | 49 | 143 | 64 |
| | 20 | 50 | 139 | 64 | 48 | 138 | 62 | 46 | 137 | 60 |
| | 19 | 46 | 133 | 60 | 44 | 132 | 58 | 43 | 131 | 57 |

Figure 34. – Cruise Performance Chart.

ASSOCIATED CONDITIONS:

| | |
|---|---|
| POWER | AS REQUIRED TO MAINTAIN 800 FT/MIN DESCENT ON APPROACH |
| FLAPS | DOWN |
| RUNWAY | PAVED, LEVEL, DRY SURFACE |
| APPROACH SPEED | IAS AS TABULATED |

NOTE: GROUND ROLL IS APPROX. 53% OF TOTAL LANDING DISTANCE OVER A 50 FT OBSTACLE.

EXAMPLE:

| | |
|---|---|
| OAT | 75 °F |
| PRESSURE ALTITUDE | 4000 FT |
| LANDING WEIGHT | 3200 LB |
| HEADWIND | 10 KNOTS |
| TOTAL LANDING DISTANCE OVER A 50 FT OBSTACLE | 1475 FT |
| GROUND ROLL (53% OF 1475) | 782 FT |
| IAS APPROACH SPEED | 87 MPH IAS |

| WEIGHT POUNDS | IAS APPROACH SPEED (ASSUMES ZERO INSTR. ERROR) | |
|---|---|---|
| | MPH | KNOTS |
| 3400 | 90 | 78 |
| 3200 | 87 | 76 |
| 3000 | 84 | 73 |
| 2800 | 81 | 70 |
| 2600 | 78 | 68 |
| 2400 | 75 | 65 |

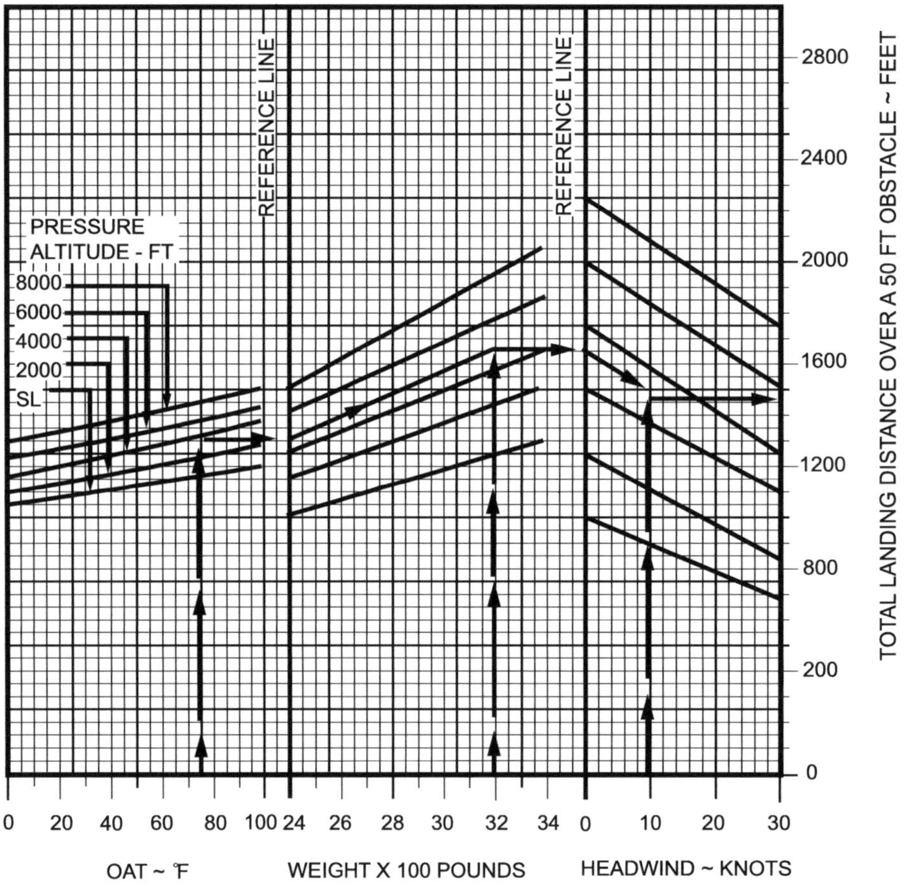


Figure 35. – Normal Landing Chart.

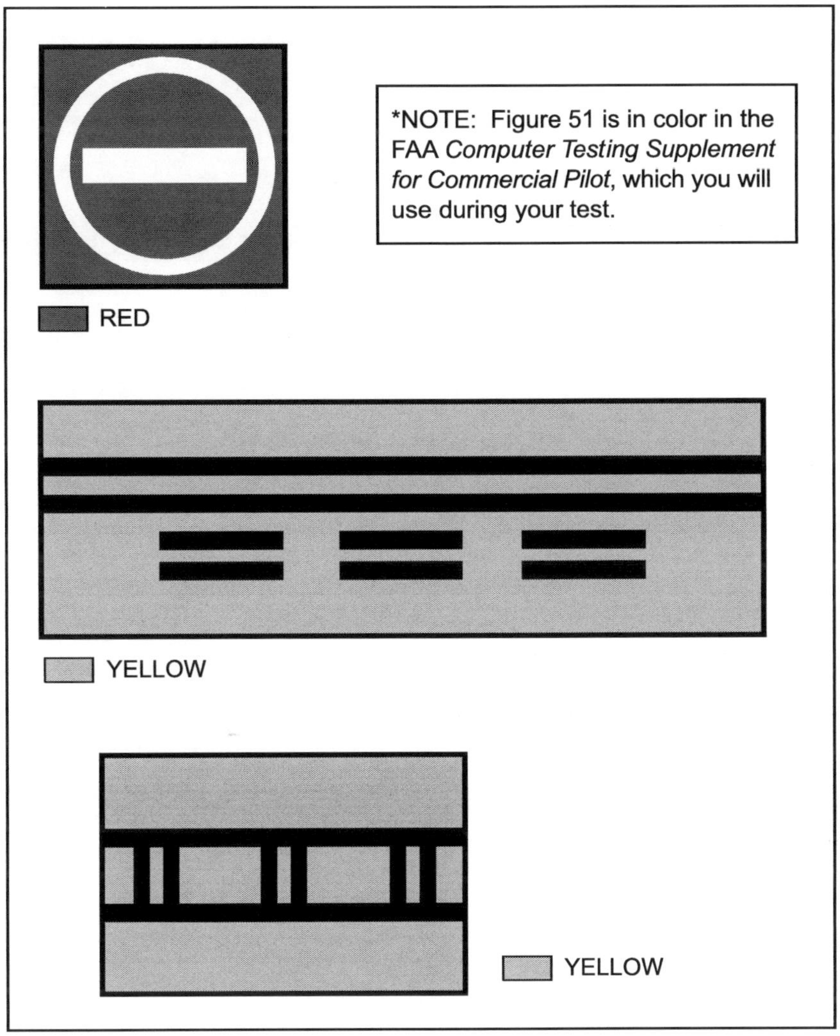


Figure 51. – Airport Signs.

Figure 64. – Airport Sign.

# APPENDIX B
# KNOWLEDGE TEST FOR
# MULTI-ENGINE LAND ADD-ON RATING

For a copy of the answer key to the *Commercial Pilot Syllabus* End-of-Course Knowledge Test, please email a request for the CPSYL End-of-Course Answer Key to the Gleim Aviation Department at aviation@gleim.com.

**Table of Contents – Multi-Engine**

# END-OF-COURSE KNOWLEDGE TEST - MULTI-ENGINE

**1.** Which of the following documents is specific to one particular airplane and its equipment, and is required to be kept current if any changes are made?

A — FAA-Approved Airplane Flight Manual.
B — Information Manual.
C — Airplane Owner's Manual.

**2.** A Minimum Equipment List (MEL) is a list of equipment that

A — Must be operative for every flight.
B — May be inoperative for some or all flights.
C — Is required for day VFR operations.

**3.** Reducing an airplane's available power by one-half reduces climb performance by

A — One-half.
B — Less than one-half.
C — More than one-half.

**4.** Asymmetric thrust is created when one engine fails on a conventional twin because the remaining engine's thrust is not distributed evenly along the

A — Longitudinal axis.
B — Lateral axis.
C — Vertical axis.

**5.** $V_{MC}$ is defined as the minimum airspeed at which directional control can be maintained with

A — Either engine inoperative and windmilling, and the other engine developing maximum power.
B — The critical engine inoperative and windmilling, and the operative engine developing maximum power.
C — The critical engine inoperative and feathered, and the operative engine developing maximum power.

**6.** During normal operations, each engine of a typical multi-engine airplane receives fuel from

A — A tank located on the opposite side of the airplane.
B — A tank located on the same side of the airplane.
C — A single main tank located in the fuselage.

**7.** Which of the following are the primary flight controls?

A — Rudder, elevator, and ailerons.
B — Flaps, trim tabs, and spoilers.
C — Throttles, propeller controls, and mixtures.

**8.** Anti- and de-icing equipment is intended primarily as a means of

A — Enabling any area of icing to be safely penetrated.
B — Removing ice build-up prior to takeoff.
C — Providing the means to escape from icing conditions.

**9.** While airborne in the vicinity of an airport with an operating control tower, a flashing red light means

A — Return for landing.
B — Airport unsafe; do not land.
C — Give way to other aircraft and continue circling.

**10.** If you are intercepted by a military or law enforcement aircraft while in distress, and you are unable to establish radio contact with the intercepting aircraft, you are expected to

A — Rock your airplane's wings.
B — Turn away from the intercepting aircraft.
C — Flash all of your aircraft's lights at irregular intervals.

**11.** Risk management is the part of the decision making process that relies on which features to reduce the risks that are associated with each flight?

A — Past experience and quick decisions.
B — The mental process of analyzing all information in a particular situation and making a timely decision on what action to take.
C — Situational awareness, problem recognition, and good judgment.

**12.** All airplane performance charts feature which item as a variable?

A — Runway gradient.
B — Barometric pressure.
C — Weight.

**13.** A multi-engine airplane's accelerate-stop distance is defined as the distance required for the airplane to accelerate to the liftoff speed ($V_{LOF}$) and, assuming the failure of one engine

A — At $V_{LOF}$, brake to a complete stop.
B — At $V_{LOF}$, continue the takeoff over a 50-ft. obstacle.
C — At $V_{YSE}$, brake to a complete stop.

**14.** The single-engine service ceiling is the maximum pressure altitude at which a multi-engine airplane is able to

A — Maintain a 100 FPM rate of climb with one engine inoperative.
B — Maintain altitude with one engine inoperative.
C — Maintain a 50 FPM rate of climb with one engine inoperative.

**15.** How can vapor lock be eliminated?

A — By using the electric fuel pump or primer to force vapor pockets out of the lines.
B — By cranking the engine with the mixture set to cut-off and the throttle fully open.
C — By using full carburetor heat at the first indication of engine roughness or loss of manifold pressure.

**16.** During the propeller feathering check, the propeller control should be placed in the feather detent just long enough for the

A — Engine to feather completely.
B — RPM to increase by 200-300 RPM.
C — RPM to decrease by 200-300 RPM.

**17.** The Practical Test Standards require that stalls be performed in a multi-engine airplane at an altitude that will allow the maneuver to be completed no lower than

A — 700 ft. AGL
B — 1,500 ft. AGL
C — 3,000 ft. AGL

**18.** When landing with a failed engine, it is better to decrease airspeed with

A — The addition of flaps.
B — A decrease in engine power.
C — Elevator trim inputs.

**19.** If an engine fails during the takeoff roll when the airspeed is below $V_{MC}$, the only correct action is to

A — Troubleshoot the failed engine while continuing the takeoff.
B — Declare an emergency while requesting priority from ATC.
C — Immediately close both throttles, maintain directional control, and apply braking as necessary.

**20.** Your first priority following the failure of one engine while airborne is to maintain

A — Altitude and heading
B — Altitude and airspeed.
C — Heading and airspeed.

**21.** Which airspeed should be maintained during an engine-inoperative approach?

A — $V_{YSE}$
B — $V_{MC}$ + 5 kt.
C — $V_{LO}$+ 5 kt.

**22.** When maneuvering with one engine inoperative, it is preferable to turn into the

A — Inoperative engine.
B — Operating engine.
C — Wind.

**23.** The density of the air DECREASES as

A — Temperature increases.
B — Humidity decreases.
C — Barometric pressure increases.

**24.** Which of the following is a type of retractable landing gear system?

A — Electro-hydraulic.
B — Turbo-compound.
C — Suction/pneumatic.

**25.** The pitot-static system provides the source for the operation of the

A — Altimeter, attitude indicator, and vertical speed indicator.
B — Altimeter, airspeed indicator, and turn coordinator.
C — Altimeter, vertical speed indicator, and airspeed indicator.

# This is to certify that

________________________________________

is enrolled in the

Federal Aviation Administration

approved Commercial Pilot Certification Course

conducted by ____________________________________

(name of school and certificate number)

__________________________ __________________________

Date of Enrollment Chief Instructor

GLEIM
KNOWLEDGE
TRANSFER
SYSTEMS®

Gleim Publications, Inc.
P.O. Box 12848
University Station
Gainesville, Florida 32604
(800) 874-5346
(352) 375-0772
(352) 375-6940 FAX
Internet: gleim.com
Email: admin@gleim.com

# This is to certify that

________________________________________

is enrolled in the

Federal Aviation Administration

approved Commercial Pilot Multi-Engine Add-on Rating Course

conducted by ________________________________________

(name of school and certificate number)

__________________________ __________________________

Date of Enrollment | Chief Instructor

GLEIM
KNOWLEDGE
TRANSFER
SYSTEMS®

Gleim Publications, Inc.
P.O. Box 12848
University Station
Gainesville, Florida 32604
(800) 874-5346
(352) 375-0772
(352) 375-6940 FAX
Internet: gleim.com
Email: admin@gleim.com

This is to certify that

_______________________________________________

has satisfactorily completed all stages, tests, and course requirements and has graduated from the
**FEDERAL AVIATION ADMINISTRATION**
approved Commercial Pilot Certification Course

conducted by ________________________________________
(name of school and certificate number)

The graduate has received _____ hours of cross-country training.

_________________________ Chief Instructor

_________________________ Date of Graduation

GLEIM KNOWLEDGE TRANSFER SYSTEMS®

Gleim Publications, Inc.
P.O. Box 12848
University Station
Gainesville, Florida 32604
(800) 874-5346
(352) 375-0772
(352) 375-6940 FAX
Internet: gleim.com
Email: admin@gleim.com

This is to certify that

________________________________________

has satisfactorily completed all stages, tests, and course requirements and has graduated from the

**FEDERAL AVIATION ADMINISTRATION**

approved Commercial Pilot Multi-Engine Add-on Rating Course

conducted by ________________________________

(name of school and certificate number)

The graduate has received _____ hours of cross-country training.

____________________ ____________________

Chief Instructor Date of Graduation

GLEIM
KNOWLEDGE
TRANSFER
SYSTEMS®

Gleim Publications, Inc.
P.O. Box 12848
University Station
Gainesville, Florida 32604
(800) 874-5346
(352) 375-0772
(352) 375-6940 FAX
Internet: gleim.com
Email: admin@gleim.com

Use the online MedXPress system (see page 2 for more information) to complete your medical application before you arrive at your aviation medical examiner's office. The FAA form below gives you an idea of what to expect when filling out the online application.

## Instructions for Completion of the Application for Airman Medical Certificate or Airman Medical and Student Pilot Certificate, FAA Form 8500-8

Applicant must fill in completely numbers 1 through 20 of the application using a ballpoint pen. Exert sufficient pressure to make legible copies. The following numbered instructions apply to the numbered headings on the application form that follows this page.
NOTICE – Intentional falsification may result in federal criminal prosecution. Intentional falsification may also result in suspension or revocation of all airman, ground instructor, and medical certificates and ratings held by you, as well as denial of this application for medical certification.

**1. APPLICATION FOR** -- Check the appropriate box.

**2. CLASS OF AIRMAN MEDICAL CERTIFICATE APPLIED FOR** -- Check the appropriate box for the class of airman medical certificate for which you are making application.

**3. FULL NAME** -- If your name has changed for any reason, list current name on the application and list any former name(s) in the EXPLANATIONS box of number 18 on the application.

**4. SOCIAL SECURITY NUMBER** -- The social security number is optional; however, its use as a unique identifier does eliminate mistakes.

**5. ADDRESS** -- Give permanent mailing address and country. Include your complete nine digit ZIP code if known. Provide your current area code and telephone number.

**6. DATE OF BIRTH** -- Specify month (MM), day (DD), and year (YYYY) in numerals; e.g., 01/31/1950. Indicate citizenship; e.g., U.S.A.

**7. COLOR OF HAIR** -- Specify as brown, black, blond, gray, or red. If bald, so state. Do not abbreviate.

**8. COLOR OF EYES** -- Specify actual eye color as brown, black, blue, hazel, gray, or green. Do not abbreviate.

**9. SEX** -- Indicate male or female.

**10. TYPE OF AIRMAN CERTIFICATE(S) YOU HOLD** -- Check applicable block(s). If "Other" is checked, provide name of certificate.

**11. OCCUPATION** -- Indicate major employment. "Pilot" will be used only for those gaining their livelihood by flying.

**12. EMPLOYER** -- Provide your employer's full name. If self-employed, so state.

**13. HAS YOUR FAA AIRMAN MEDICAL CERTIFICATE EVER BEEN DENIED, SUSPENDED, OR REVOKED** -- If "yes" is checked, give month and year of action in numerals.

**14. TOTAL PILOT TIME TO DATE** -- Give total number of civilian flight hours. Indicate whether logged or estimated. Abbreviate as Log. or Est.

**15. TOTAL PILOT TIME PAST 6 MONTHS** -- Give number of civilian flight hours in the 6-month period immediately preceding date of this application. Indicate whether logged or estimated. Abbreviate as Log. or Est.

**16. MONTH AND YEAR OF LAST FAA MEDICAL EXAMINATION** -- Give month and year in numerals. If none, so state.

**17a. DO YOU CURRENTLY USE ANY MEDICATION (Prescription or Nonprescription)** -- Check "yes" or "no." If "yes" is checked, give name of medication(s) and indicate if the medication was listed in a previous FAA medical examination. See **NOTE** below.

**17b**. Indicate whether you use near vision contact lens(es) while flying.

**18. MEDICAL HISTORY** -- Each item under this heading must be checked either "yes" or "no." You must answer "yes" for every condition you have ever been diagnosed with, had, or presently have and describe the condition and approximate date in the EXPLANATIONS block.

If information has been reported on a previous application for airman medical certificate and there has been no change in your condition, you may note "PREVIOUSLY REPORTED, NO CHANGE" in the EXPLANATIONS box, but you must still check "yes" to the condition. Do not report occasional common illnesses such as colds or sore throats.

"Substance dependence" is defined by any of the following: increased tolerance; withdrawal symptoms; impaired control of use; or continued use despite damage to health or impairment of social, personal, or occupational functioning. "Substance abuse" includes the following: use of an illegal substance; use of a substance or substances in situations in which such use is physically hazardous; or misuse of a substance when such misuse has impaired health or social or occupational functioning. "Substances" include alcohol, PCP, marijuana, cocaine, amphetamines, barbiturates, opiates, and other psychoactive chemicals.

Conviction and/or Administrative Action History -- Letter (v) of this subheading asks if you have ever been: (1) convicted (which may include paying a fine, or forfeiting bond or collateral) of an offense involving driving while intoxicated by, while impaired by, or while under the influence of alcohol or a drug; or (2) convicted or subject to an administrative action by a state or other jurisdiction for an offense for which your license was denied, suspended, cancelled, or revoked or which resulted in attendance at an educational or rehabilitation program. Individual traffic convictions are not required to be reported if they did not involve: alcohol or a drug; suspension, revocation, cancellation, or denial of driving privileges; or attendance at an educational or rehabilitation program. If "yes" is checked, a description of the conviction(s) and/or administrative action(s) must be given in the EXPLANATIONS box. The description must include: (1) the alcohol or drug offense for which you were convicted or the type of administrative action involved (e.g., attendance at an alcohol treatment program in lieu of conviction; license denial, suspension, cancellation, or revocation for refusal to be tested; educational safe driving program for multiple speeding convictions; etc.); (2) the name of the state or other jurisdiction involved; and (3) the date of the conviction and/or administrative action. The FAA may check state motor vehicle driver licensing records to verify your responses. Letter (w) of this subheading asks if you have ever had any other (nontraffic) convictions (e.g., assault, battery, public intoxication, robbery, etc.). If so, name the charge for which you were convicted and the date of conviction in the EXPLANATIONS box. See **NOTE** below.

**19. VISITS TO HEALTH PROFESSIONAL WITHIN LAST 3 YEARS** – List all visits in the last 3 years to a physician, physician assistant, nurse practitioner, psychologist, clinical social worker, or substance abuse specialist for treatment, examination, or medical/mental evaluation. List visits for counseling only if related to a personal substance abuse or psychiatric condition. Give date, name, address, and type of health professional consulted and briefly state reason for consultation. Multiple visits to one health professional for the same condition may be aggregated on one line. Routine dental, eye, and FAA periodic medical examinations and consultations with your employer-sponsored employee assistance program (EAP) may be excluded unless the consultations were for your substance abuse or unless the consultations resulted in referral for psychiatric evaluation or treatment. See **NOTE** below.

**20. APPLICANT'S DECLARATION** -- Two declarations are contained under this heading. The first authorizes the National Driver Register to release adverse driver history information, if any, about the applicant to the FAA. The second certifies the completeness and truthfulness of the applicant's responses on the medical application. The declaration section must be signed and dated by the applicant after the applicant has read it.

**NOTE:** If more space is required to respond to "yes" answers for numbers 17, 18, or 19, use a plain sheet of paper bearing the information, your signature, and the date signed.

**Applicant -- Please Tear Off This Sheet After Completing The Application Form.**

FAA Form 8500-8 (3-99) Supersedes Previous Edition

NSN-0052-00-670-6002

**Applicant Must Complete ALL 20 Items (Except For Shaded Areas) PLEASE PRINT** Form Approved OMB NO. 2120-0034

Copy of FAA Form 8500-8 (Medical Certificate) or FAA Form B420-2 Medical/Student Pilot Certificate) Issued. **FF- 1953420**

**MEDICAL CERTIFICATE ______________ CLASS AND STUDENT PILOT CERTIFICATE**

This certifies that (*Full name and address*):

| Date of Birth | Height | Weight | Hair | Eyes | Sex |
|---|---|---|---|---|---|
| | | | | | |

has met the medical standards prescribed in Part 67, Federal Aviation Regulations, for this class of Medical Certificate

Limitations

| Date of Examination | Examiner's Designation No. |
|---|---|
| | |

Examiner: Signature

Typed Name

**AIRMAN'S SIGNATURE**

**1. Application For:** ☐ Airman Medical Certificate ☐ Airman Medical and Student Pilot Certificate

**2. Class of Medical Certificate Applied For:** ☐ 1st ☐ 2nd ☐ 3rd

**3. Last Name** **First Name** **Middle Name**

**4. Social Security Number** — —

**5. Address** Telephone Number ( ) -

Number/Street

City State/Country Zip Code

**6. Date of Birth** ______ MM / DD / YYYY **Citizenship** ______

**7. Color of Hair** **8. Color of Eyes** **9. Sex**

**10. Type of Airman Certificate(s) You Hold:**

☐ None ☐ ATC Specialist ☐ Flight Instructor ☐ Recreational

☐ Airline Transport ☐ Flight Engineer ☐ Private ☐ Other

☐ Commercial ☐ Flight Navigator ☐ Student ______

**11. Occupation** **12. Employer**

**13. Has Your FAA Airman Medical Certificate Ever Been Denied, Suspended, or Revoked?**
☐ Yes ☐ No If yes, give date ______ MM / DD / YYYY

**Total Pilot Time** (Civilian Only)
**14. To Date** **15. Past 6 months**

**16. Date of Last FAA Medical Application** ______ MM / DD / YYYY ☐ No Prior Application

**17a. Do You Currently Use Any Medication (Prescription or Nonprescription)?**
☐ No ☐ Yes (If yes, below list medication(s) used and check appropriate box) Previously Reported Yes No

______ ☐ ☐
______ ☐ ☐
______ ☐ ☐

(If more space is required, see 17.a. on the instruction sheet).

**17b. Do You Ever Use Near Vision Contact Lens(es) While Flying?** ☐ Yes ☐ No

**18. Medical History -** HAVE YOU EVER IN YOUR LIFE BEEN DIAGNOSED WITH, HAD, OR DO YOU PRESENTLY HAVE ANY OF THE FOLLOWING? Answer "yes" or "no" for every condition listed below. In the EXPLANATIONS box below, you may note "PREVIOUSLY REPORTED, NO CHANGE" only if the explanation of the condition was reported on a previous application for an airman medical certificate and there has been no change in your condition. **See Instructions Page**

| Yes | No | Condition | Yes | No | Condition | Yes | No | Condition | Yes | No | Condition |
|---|---|---|---|---|---|---|---|---|---|---|---|
| a. ☐ | ☐ | Frequent or severe headaches | g. ☐ | ☐ | Heart or vascular trouble | m. ☐ | ☐ | Mental disorders of any sort: depression, anxiety, etc. | r. ☐ | ☐ | Military medical discharge |
| b. ☐ | ☐ | Dizziness or fainting spell | h. ☐ | ☐ | High or low blood pressure | n. ☐ | ☐ | Substance dependence or failed a drug test ever, or substance abuse or use of illegal substance in the last 2 years. | s. ☐ | ☐ | Medical rejection by military service |
| c. ☐ | ☐ | Unconsciousness for any reason | i. ☐ | ☐ | Stomach, liver, or intestinal trouble | | | | t. ☐ | ☐ | Rejection for life or health insurance |
| d. ☐ | ☐ | Eye or vision trouble except glasses | j. ☐ | ☐ | Kidney stone or blood in urine | o. ☐ | ☐ | Alcohol dependence or abuse | u. ☐ | ☐ | Admission to hospital |
| e. ☐ | ☐ | Hay fever or allergy | k. ☐ | ☐ | Diabetes | p. ☐ | ☐ | Suicide attempt | x. ☐ | ☐ | Other illness, disability, or surgery |
| f. ☐ | ☐ | Asthma or lung disease | l. ☐ | ☐ | Neurological disorders: epilepsy, seizures, stroke, paralysis, etc. | q. ☐ | ☐ | Motion sickness requiring medication | | | |

**Conviction and/or Administrative Action History -- See Instructions Page**

| Yes | No | | Yes | No | |
|---|---|---|---|---|---|
| v. ☐ | ☐ | History of (1) any conviction(s) involving driving while intoxicated by, while impaired by, or while under the influence of alcohol or a drug; or (2) history of any conviction(s) or administrative action(s) involving an offense(s) which resulted in the denial, suspension, cancellation, or revocation of driving privileges or which resulted in attendance at an educational or a rehabilitation program. | w. ☐ | ☐ | History of nontraffic conviction(s) (misdemeanors or felonies). |

**Explanations:** See Instructions Page

FOR FAA USE Review Action Codes

**19. Visits to Health Professional Within Last 3 Years.** ☐ Yes (explain below) ☐ No **See Instructions Page**

| Date | Name, Address, and Type of Health Professional Consulted | Reason |
|---|---|---|
| | | |
| | | |
| | | |

-- NOTICE --
Whoever in any matter within the jurisdiction of any department or agency of the United States knowingly and willfully falsifies, conceals or covers up by any trick, scheme, or device a material fact, or who makes any false, fictitious, or fraudulent statements or representations, or entry, may be fined up to $250,000 or imprisoned not more than 5 years, or both, (18 U.S. Code Secs. 1001; 3571).

**20. Applicant's National Driver Register and Certifying Declarations**

I hereby authorize the National Driver Register (NDR), through a designated State Department of Motor Vehicles, to furnish to the FAA information pertaining to my driving record. This consent constitutes authorization for a single access to the information contained in the NDR to verify information provided in this application. Upon my request, the FAA shall make the information received from the NDR, if any, available for my review and written comment. Authority: 23 U.S. Code 401, Note.

**NOTE: All persons using this form must sign it. NDR consent, however, does not apply unless this form is used as an application for Medical Certificate or Medical Certificate and Student Pilot Certificate.**

I hereby certify that all statements and answers provided by me on this application form are complete and true to the best of my knowledge, and I agree that they are to be considered part of the basis for issuance of any FAA certificate to me. I have also read and understand the Privacy Act statement that accompanies this form.

Signature of Applicant Date ______ MM/DD/YYYY

**FAA Form 8500-8 (3-00) Supersedes Previous Edition** **NSN-0052-00-670-6002**

**NOTE: FAA/Original Copy of the Report of Medical Examination Must Be TYPED.**

## REPORT OF MEDICAL EXAMINATION

| 21. Height (inches) | 22. Weight (pounds) | 23. Statement of Demonstrated Ability (SODA) ☐ Yes ☐ No Defect Noted: | 24. SODA Serial Number |
|---|---|---|---|
| | | | |

| CHECK EACH ITEM IN APPROPRIATE COLUMN | Normal | Abnormal | CHECK EACH ITEM IN APPROPRIATE COLUMN | Normal | Abnormal |
|---|---|---|---|---|---|
| 25. Head, face, neck, and scalp | | | 37. Vascular system (Pulse, amplitude and character; arms, legs, others) | | |
| 26. Nose | | | 38. Abdomen and viscera (Including hernia) | | |
| 27. Sinuses | | | 39. Anus (Not including digital examination) | | |
| 28. Mouth and throat | | | 40. Skin | | |
| 29. Ears, general (Internal and external canals; Hearing under item 49) | | | 41. G-U system (Not including pelvic examination) | | |
| 30. Ear Drums (Perforation) | | | 42. Upper and lower extremities (Strength and range of motion) | | |
| 31. Eyes, general (Vision under items 50 to 54) | | | 43. Spine, other musculoskeletal | | |
| 32. Ophthalmoscopic | | | 44. Identifying body marks, scars, tattoos (Size and location) | | |
| 33. Pupils (Equality and reaction) | | | 45. Lymphatics | | |
| 34. Ocular motility (Associated parallel movement, nystagmus) | | | 46. Neurologic (Tendon reflexes, equilibrium, senses, cranial nerves, coordination, etc.) | | |
| 35. Lungs and chest (Not including breasts examination) | | | 47. Psychiatric (Appearance, behavior, mood, communication, and memory) | | |
| 36. Heart (Precordial activity, rhythm, sounds, and murmurs) | | | 48. General systemic | | |

**NOTES:** Describe every abnormality in detail. Enter applicable item number before each comment. Use additional sheets if necessary and attach to this form.

| 49. Hearing | Record Audiometric Speech Discrimination Score Below | | Right Ear | | | | | Left Ear | | | | |
|---|---|---|---|---|---|---|---|---|---|---|---|---|
| Conversational Voice Test at 6 Feet ☐ Pass ☐ Fail | | Audiometer Threshold in Decibels | 500 | 1000 | 2000 | 3000 | 4000 | 500 | 1000 | 2000 | 3000 | 4000 |
| | | | | | | | | | | | | |

| 50. Distant Vision | 51.a. Near Vision | 51.b. Intermediate Vision - 32 Inches | 52. Color Vision |
|---|---|---|---|
| Right 20/ Corrected to 20/<br>Left 20/ Corrected to 20/<br>Both 20/ Corrected to 20/ | Right 20/ Corrected to 20/<br>Left 20/ Corrected to 20/<br>Both 20/ Corrected to 20/ | Right 20/ Corrected to 20/<br>Left 20/ Corrected to 20/<br>Both 20/ Corrected to 20/ | ☐ Pass<br>☐ Fail |

| 53. Field of Vision | 54. Heterophoria 20' (in prism diopters) | Esophoria | Exophoria | Right Hyperphoria | Left Hyperphoria |
|---|---|---|---|---|---|
| ☐ Normal ☐ Abnormal | | | | | |

| 55. Blood Pressure (Sitting, mm of Mercury) | Systolic | Diastolic | 56. Pulse (Resting) | 57. Urinalysis (if abnormal, give results) ☐ Normal ☐ Abnormal | Albumin | Sugar | 58. ECG (Date) MM \| DD \| YYYY |
|---|---|---|---|---|---|---|---|
| | / | | | | | | |

**59. Other Tests Given**

**60. Comments on History and Findings:** AME shall comment on all "YES" answers in the Medical History section and for abnormal findings of the examination. (Attach all consultation reports, ECGs, X-rays, etc. to this report before mailing.)

| FOR FAA USE |
|---|
| Pathology Codes: |
| Coded By: |
| Clerical Reject |

**Significant Medical History** ☐ YES ☐ NO **Abnormal Physical Findings** ☐ YES ☐ NO

**61. Applicant's Name**

**62. Has Been Issued --** ☐ Medical Certificate ☐ Medical & Student Pilot Certificate
☐ **No Certificate Issued --** Deferred for Further Evaluation
☐ **Has Been Denied --** Letter of Denial Issued (Copy Attached)

**63. Disqualifying Defects** (List by item number)

**64. Medical Examiner's Declaration --** I hereby certify that I have personally reviewed the medical history and personally examined the applicant named on this medical examination report. This report with any attachment embodies my findings completely and correctly.

| Date of Examination | Aviation Medical Examiner's Name | Aviation Medical Examiner's Signature |
|---|---|---|
| MM \| DD \| YYYY | Street Address | |
| | | AME Serial Number |
| | City State Zip Code | AME Telephone ( ) |

**FAA Form 8500-8** (7-92) Supersedes Previous Editions

Please forward your suggestions, corrections, and comments concerning typographical errors, etc., to **Irvin N. Gleim • c/o Gleim Publications, Inc. • P.O. Box 12848 • University Station • Gainesville, Florida • 32604.** Please include your name and address so we can properly thank you for your interest.

1. ______________________________

2. ______________________________

3. ______________________________

4. ______________________________

5. ______________________________

6. ______________________________

7. ______________________________

Remember, for superior service: Email, mail, or fax questions about our materials.
Telephone questions about orders, prices, shipments, or payments.

Name: ______________________________

Address: ______________________________

City/State/Zip: ______________________________

Telephone: Home: __________ Work: __________ Fax: __________

Email: ______________________________